Introduction to Thermodynamics of Mechanical Fatigue

Introduction to Thermodynamics of Mechanical Fatigue

Michael M. Khonsari • Mehdi Amiri

CRC Press
Taylor & Francis Group
Boca Raton   London   New York

CRC Press is an imprint of the
Taylor & Francis Group, an **informa** business

CRC Press
Taylor & Francis Group
6000 Broken Sound Parkway NW, Suite 300
Boca Raton, FL 33487-2742

First issued in paperback 2017

Version Date: 20120822

ISBN 13: 978-1-4665-1179-8 (hbk)
ISBN 13: 978-1-138-07192-6 (pbk)

Library of Congress Cataloging-in-Publication Data

Khonsari, Michael M.
  Introduction to thermodynamics of mechanical fatigue / Michael M. Khonsari, Mehdi Amiri.
    p. cm.
  Includes bibliographical references and index.
  ISBN 978-1-4665-1179-8 (hardback)
  1. Materials--Fatigue. 2. Materials--Thermal properties. I. Amiri, Mehdi. II. Title.

TA418.38.K45 2012
620.1'126--dc23   2012027082

Visit the Taylor & Francis Web site at
http://www.taylorandfrancis.com

and the CRC Press Web site at
http://www.crcpress.com

Dedicated to

_____

Karen, Maxwell,
Milton, Mason Khonsari, and
to the memory of my father (MMK),
and to Hassan Amiri

Contents

Preface

The subject of fatigue degradation, and the methodologies for its treatment, spans multitudes of scientific disciplines, ranging from engineering to materials science and mechanics to mathematics.

Fatigue is probabilistic in nature. For example, fatigue tests performed on the same material subjected to the same operating conditions can yield different results in terms of the number of cycles that the system can withstand before failure occurs. Such uncertainties affect the system design, structural integrity, and operational reliability. Yet the majority of available methods for prediction of fatigue failure—cumulative damage models, cyclic plastic energy hypothesis, crack propagation rate models, and empirically-derived relationships based on the curve fitting of the limited laboratory data—require many unknown input parameters that must be experimentally determined.

There are other complications. All of the above-mentioned methods concentrate on very specific types of loading and single fatigue modes, that is, bending, torsion, or tension-compression. In practice, however, fatigue involves simultaneous interaction of multimode processes. Further, the variability in the duty cycle in practical applications may render many of these existing methods incapable of reliable prediction. It is, therefore, no surprise that the application of these theories often leads to many uncertainties in the design. Further, their use and execution in practice requires one to implement large factors of safety, often leading to gross overdesigns that waste resources and cost more.

In reality, the science base that underlies modeling and analysis of fatigue processes has remained substantially unchanged for decades, leaving a significant gap between the available technology and the science that effectively captures the dynamics of degradation. The premise of this textbook is that fatigue is a dissipative process and must obey the laws of thermodynamics. In general, it can be hypothesized that the degradation of machinery components is a consequence of irreversible thermodynamic processes that disorder a component, and that degradation is a time-dependent phenomenon with increasing disorder. This suggests that entropy—a fundamental parameter in thermodynamics that characterizes disorder—offers a natural measure of component degradation.

While an entropic approach to problems involving degradation is gaining momentum, its practical applications have not yet become widespread. This concept offers new and exciting research in the field of fatigue fracture analysis for years to come. We hope this introduction to the treatment of fatigue via the principles of thermodynamics serves as a useful contribution to the science of degradation.

MMK and MA
Baton Rouge, Louisiana

About the Authors

Michael M. Khonsari is the holder of the Dow Chemical Endowed Chair and professor of the Mechanical Engineering Department at Louisiana State University, where he directs the Center for Rotating Machinery. Professor Khonsari is a fellow of the American Society of Mechanical Engineers (ASME), the Society of Tribologists and Lubrication Engineers (STLE), and the American Association for the Advancement of Science (AAAS). He holds several patents and has authored two books and over 200 archival journal articles. He is currently the editor of the *ASME Journal of Tribology*.

Mehdi Amiri earned his PhD in mechanical engineering from Louisiana State University, where he is currently a research associate in the Center for Rotating Machinery. His area of research is in the field of fatigue and fracture analysis. He holds one patent and has authored several journal publications. His research interests include thermal/fluid mechanics, thermodynamics, tribology, and failure analysis.

Acknowledgments

The authors gratefully acknowledge the assistance of their colleagues in the LSU Center for Rotating Machinery. In particular, the authors thank Dr. Mehdi Naderi, Md Liakat Ali, and Paul Williams for suggesting numerous improvements. The authors specially thank Dr. Mehdi Naderi for very fruitful discussions during the course of the research and development associated with this subject matter.

1 Introduction to Mechanical Degradation Processes

In this chapter, we review some of the commonly observed failure modes in structures and machinery components. We are particularly interested in the science of degradation to understand the nature of progressive change in the integrity of material properties to the point that the component can no longer function in the intended manner for which it was designed. Mechanical failure does not necessarily mean fracture. Any drastic change in shape, size, or material properties that renders a component incapable of performing its intended function may be considered a failure (Collins 1993). To describe the evolution of failure, therefore, it is more appropriate to adopt the terminology of degradation or aging instead of failure.

There are two factors that play key roles in all degradation processes. They are forces (stresses) and time. Here, the term *stress* is not limited to mechanical stress. It can be induced mechanically, thermally, or chemically. A brief description of pertinent degradation processes follows.

1.1 FATIGUE

Fatigue is a progressive decrease of material strength as a result of the application of cyclic mechanical load or cyclic deformation. Fatigue failure can occur suddenly without any noticeable warning and often catastrophically. It begins by formation of microcracks that continue to grow with repeated application of load and concomitantly reduces the material's residual strength until it becomes so low that the failure of the structure becomes imminent (Broek 1982).

Fatigue is one of the most predominate modes that cause failure in a diverse array of man-made components and natural systems (Stephens et al. 2000). Frequent occurrence of fatigue damage in structures and devices has furnished the incentive for investigating its mechanism for more than 160 years. Clearly, preventing failure through reliable lifetime prediction methodologies would have a major societal impact in terms of both economics and safety. According to the National Bureau of Standards, the costs associated with material fractures for a single year (1978) in the Unites States was $119 billion dollars per year (1982 dollars) or 4% of the Gross National Product. The report contends that approximately 1/3 ($35 billion per year) could be saved through the use of available technology and that another 1/4 ($28 billion per year) could be saved through fracture-related research (Reed, Smith, and Christ 1983).

Aside from economic considerations, environmental and safety aspects of fatigue failure are also crucially important to the society. Studies by Das, Cheng, and Murray (2007) show that geotechnical movements impose large displacement on steel pipelines that transport petroleum products, natural gas, and so forth. These displacements are known to cause low-cycle fatigue fracture with serious environmental consequences.

FIGURE 1.1   Fuselage fatigue failure on the Aloha Airlines Boeing 737 after 89,090 flight cycles. (From Vogelesang, L.B. and Vlot, A., *J. Mater. Proc. Tech.*, 103, 1–5, 2000. With permission.)

A study by Bhaumik, Sujata, and Ventataswamy (2008) revealed that about 60% of the failure of aircraft components in service is due to fatigue. A horrific example of fatigue failure is the Aloha Airline aircraft accident that occurred on April 28, 1988 (Figure 1.1). According to the report of the National Transportation Safety Board (NTSB Number: AAR-89-03), fatigue was the main cause of failure since the plane had experienced an unusually high number of fatigue cycles due to frequent takeoff-landing cycles among islands in Hawaii. As a result, while the plane was in flight, approximately 18 ft of the cabin was ripped away at the altitude of 24,000 ft.

1.2   FRACTURE

Fracture is a failure mode that occurs in both brittle and ductile materials. Brittle fracture is accompanied by the break up of a material into two pieces when subjected to monotonic load; imagine twisting or bending a piece of chalk and its sudden fracture into two pieces without any warning. In contrast, fracture in ductile materials is preceded by plastic deformation and a measurable delay before the onset of fracture. Particularly susceptible are flaws and stress concentration sites in material where the strength of material reduces significantly and structure or component fails at low stresses (Broek 1982). An example of brittle fracture is the catastrophic fracture of the MV Kurdistan tanker on March 15, 1979, while carrying 30,000 tons of fuel oil (Figure 1.2). According to Garwood (1997), brittle fracture was initiated in the port bilge keel weld and propagated into the ship's structure, causing its breakage in two. A thorough review of catastrophic failures due to fracture of metals from the last 200 years is given by Anderson (1969).

1.3   WEAR

Wear is defined as the progressive removal of material from the surface of bodies in contact as a result of relative motion. Wear is predominately a gradual process and does not necessarily cause a sudden catastrophic failure. As a result of wear, a component may suffer from

FIGURE 1.2 MV Kurdistan after fracture. (From Garwood, S.J., *Eng. Failure Anal.*, 4, 3–24, 1997. With permission.)

surface damage and dimensional changes that may give rise to misalignment, vibration, and increase in local stress level. Wear can also be a serious cause of energy loss and material degradation, thus contributing to reduction of efficiency and power.

The mode of contact—sliding, rolling, oscillating, impacting, or fretting—governs the nature of wear in a tribosystem. While specific classifications exist, one can broadly categorize wear in terms of the responsible mechanical, thermal, and chemical degradation process, depending on the cause of surface damage (Stachowiak 2005). For instance, mechanical wear mainly occurs as a result of deformation and damage of the interfacial surfaces due to the applied load. At microscale, wear mechanism is governed by plastic deformation of asperities and fracture of asperity junctions, resulting in material removal by what is known as adhesive wear. In contrast, in abrasive wear, damage occurs by the cutting action of a hard protuberance at the interface where it leaves a scratch, a groove, or an indentation on the surface (Bayer 1994). Figure 1.3 shows examples of typical adhesive and abrasive wear. The reader interested in detailed discussion on wear analysis is referred to, for example, Stachowiak (2005), Bayer (1994), and Collins (1993).

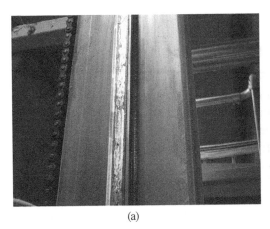

(a)   (b)

FIGURE 1.3 (a) Adhesive wear on a rail of column lift. (From Olsson, M., *Wear*, 270, 720–724, 2011. With permission.) (b) Abrasive wear of a 200 μm radius diamond against Al 6061-T6. (From Kobrick, R.L., Klaus, D.M. and Street, K.W., *Wear*, 270, 650–657, 2011. With permission.)

Fretting fatigue cracks formed at these locations.

FIGURE 1.4 A gas turbine disc/shaft connection is damaged by fretting. (From Hoeppner, D.W., *Tribology Int.*, 39, 1271–1276, 2006. With permission.)

1.4 FRETTING

Fretting is a repetitive, oscillatory, small-amplitude motion between two bodies that degrades the contacting surfaces due to repeated shear forces. Fretting commonly occurs in structures and machine elements that are clamped together, or those that are in contact while undergoing reciprocating slippage over surfaces. It is commonly observed in contacting assemblies such as bearings, mechanically fastened joints, turbine blade pair, and artificial hip joints. Common fretting modes are fretting fatigue and fretting wear. Fretting fatigue is, however, known to be the root cause of numerous catastrophic failures in the industry. Many fascinating examples of fretting failures in helicopters, fixed wing aircraft, trains, ships, automobiles, trucks and buses, farm machinery, engines, construction equipment, orthopedic implants, artificial hearts, rocket motor cases, wire ropes, and so forth, are reported by Hoeppner (1992). Figure 1.4 shows a typical fretting fatigue failure that occurred in a gas turbine disc/shaft connection (Hoeppner 2006). According to Hoeppner, this failure had led to a crash and loss of life.

1.5 BRINELLING AND FALSE BRINELLING

Brinelling occurs as a result of static overloading between two surfaces in contact that causes local yielding and permanent indentation. Brinelling damage is frequently observed in idle rolling element bearings under heavy static loading that tends to produce surface discontinuity in the raceway. These discontinuities cause premature failure when the bearing is put in operation. The terminology is taken from Brinell hardness, as the surface indentations resemble those observed during hardness testing. While brinelling damage in lightly-loaded bearings is rare, a form of *false brinelling* is frequently observed when dealing with stationary contacting surfaces that undergo vibration. According to Schoen et al. (1995) "even though lightly loaded bearings are less susceptible, false brinelling still happens and has even occurred during the transportation of uninstalled bearings." Figure 1.5 shows a typical false brinelling of a rolling bearing of the bleed system valve exposed to high vibrations induced by an aircraft engine (Massi et al. 2010).

FIGURE 1.5   False brinelling due to high vibration of a rolling bearing. (From Massi, F., Rocchi, J., Culla, A., and Berthier, Y., *Mech. Sys. Signal Proc.*, 24, 1068–1080, 2010. With permission.)

1.6   CORROSION

Corrosion is a type of degradation that occurs due to chemical or electrochemical reaction with environment. The root cause of corrosion damage is destructive oxidation, which distinguishes it from wear, fretting, and brinelling failures that are associated with mechanical loading. Corrosion mainly occurs due to long-term exposure of metal to a corrosive environment with often catastrophic consequences as a result of premature failure of bridges, pipelines, tanks, ships, marine structures, metal parts of motor vehicles, aircraft airframes and nuclear reactors. According to a recent study by Koch et al. (2002), the costs associated with material corrosion for a single year in the United States were over $275 billion, or 3.1% of the Gross Domestic Product (GDP). The report also contends that about 25 to 30% of this total could have been saved through effective application of currently available technologies that are designed to combat corrosion. Aside from economic considerations, environmental and safety aspects of corrosion are also crucially important to the society. Figure 1.6 shows a photograph of the Silver Bridge over the Ohio River, which collapsed

FIGURE 1.6   Silver Bridge collapse as a result of stress and corrosion. (From *The Herald-Dispatch* newspaper, December 14, 2011. With permission.)

on December 15, 1967, resulting in death of 46 people (McCafferty 2010). Combined effect of corrosion and stress was reported to be the cause of failure. The reader interested in a detailed discussion on corrosion analysis is referred to reports by Schweitzer (2010), Revie and Uhlig (2008), and Davis (2000).

1.7 CREEP

Creep is an excessive plastic deformation of a material under constant stress at elevated temperature over a period of time. Elevated temperature is often greater than roughly half of melting point of the material (Kassner 2009). Creep failure is not necessarily rupture, and it occurs when progressive plastic deformation at elevated temperatures changes the physical dimensions of a component in a manner that renders it incapable of functioning properly. Creep damage basically results from the initiation, growth, and coalescence of microcavities and microcracks. It is commonly observed in machinery components that operate at high temperatures such as aeroengines, components of power generation equipment, and chemical refineries and high-temperature pipelines. An example of catastrophic creep failure is the collapse of a spray drier at the Western Platinum Mine metallurgical plant after 20 years of being in service (Carter 1998). Results of ultrasonic non-destructive testing (NDT) showed that the collapse of the structure was due to excessive creep deformation (see Figure 1.7). The interested reader is referred to the work of Lemaitre and Desmorat (2005), Betten (2005), and Naumenko and Altenbach (2007) for pertinent discussion on this subject.

1.8 THERMAL SHOCK

Thermal shock failure occurs when a body is exposed to a rapid change in the environmental temperature (Hesselman 1969). Thermal stresses induced by shock temperature difference are often so pronounced that they cause fracture failure. Thermal shock failure is commonly observed in ceramics that are widely used in high temperature applications such as gas turbines and propulsion systems of aerospace planes (Schneider 1993). Figure 1.8

FIGURE 1.7 Photograph of collapsed drier due to creep. (From Carter, P., *Eng. Failure Anal.*, 5, 143–147, 1998. With permission.)

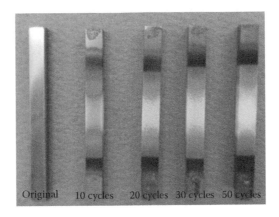

Original   10 cycles   20 cycles   30 cycles   50 cycles

FIGURE 1.8 Photograph of ZrB_2–SiC–ZrC ceramic undergoes thermal shock. (From Qi, F., Meng, S., and Guo, H., *Mater. Design*, 35, 133–137, 2012. With permission.)

shows an image of a ZrB_2–SiC–ZrC ceramic specimen before and after 10, 20, 30, and 50 thermal cycles (Qi, Meng, and Guo 2012). In each thermal cycle, temperature changes from ambient to 1800°C in 5 s which results in surface oxidation (see Figure 1.8). Qi, Meng, and Guo (2012) observed a significant damage in the form of through holes and bubbles on the surface of the specimen after 50 cycles.

1.9 IMPACT

Impact failure occurs when a structure or machinery component is hit by the collision of a foreign object(s). Impact failure mechanisms can vary depending on the kinetic energy of the collision. High-energy impact often yields a complete penetration with observable surface damage or separation of the member, while low-energy impact creates internal damages that cannot be visually detected. These internal damages can cause substantial degradation in important mechanical properties including strength and stiffness. Impact damage is the major cause of failure in structures and components made of composite laminates. In recent years, fiber reinforced composites have been widely used in numerous applications mainly due to their potential for designing light-weight structures such as air-crafts, armor systems, and combat vehicles. It is well known that composite structures are extremely susceptible to delamination and failure when subjected to external impact load-ing due to their relatively weak tolerance to localized impact. For example, airplane struc-tures are frequently subjected to impact damages caused by dropped tools, runway stones, or large hails. Therefore, it is of paramount importance to assess residual strength and per-formance of these materials under impact loading. A common example of impact damage with enormous safety implication is the failure of receptacle seat belts. Seat belt receptacles are susceptible to damage due to the occurrence of repeated, low-energy impact over a period of time. Figure 1.9 shows a typical example of seat belt damage induced by impact. Comprehensive reviews on impact damage can be found, for example, in Richardson and Wisheart (1996), Cantwell and Morton (1991), Degrieck and Paepegem (2001), and in a book by Abrate (2005).

Table 1.1 shows a summary of degradation processes, root cause, and examples of dif-ferent types of failure.

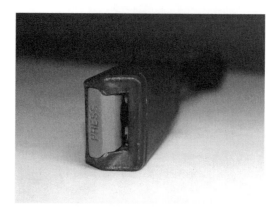

FIGURE 1.9   Photograph of a Takata-manufactured receptacle seat belt damaged by impact. (From Henshaw, J.M., Wood, V., and Hall, A.C., *Eng. Failure Anal.* 6, 13–25, 1999. With permission.)

TABLE 1.1
An Overview of Common Degradation Processes

Process	Cause of Failure	Examples
Fatigue	Repeated application of mechanical stress or strain	Aircraft structures, bridges, railroad structures, rotating shafts, turbine blades
Fracture	Monotonic mechanical load	Ship structures, welded joints, composite structures
Wear	Relative motion of contacting surfaces	Seals, brakes, bearing elements, clutches, gears
Fretting	Oscillatory small amplitude shear stress (slip) at contacting surfaces	Wire ropes, fastened joints, turbine blades
False Brinelling	Vibration of two stationary contacting elements	Stationary bearings, induction motors
Corrosion	Chemical or electrochemical reaction with environment	Bridges, pipelines, tanks, marine structures, ships
Creep	Excessive plastic deformation at elevated temperatures	Aeroengines, high-temperature pipelines, chemical refineries
Thermal Shock	Excessive temperature gradient	Ceramics used in turbines and propulsion systems
Impact	Shock loading due to collision of two bodies	Aircrafts, armor systems, combat vehicles

REFERENCES

Abrate, S. 2005. *Impact on Composite Structures*. Cambridge, UK: Cambridge University Press.
Anderson, W.E. 1969. *An Engineer Views Brittle Fracture History*. Boeing rept.
Bayer, R.G. 1994. *Mechanical Wear Prediction and Prevention*. New York: Marcel Dekker, Inc.
Betten, J. 2005. *Creep Mechanics*. 2nd ed. Berlin Heidelberg: Springer-Verlag.
Bhaumik, S.K., Sujata, M., and Ventataswamy, M.A. 2008. Fatigue failure of aircraft components. *Eng. Failure Anal.* 15, 675–894.
Broek, D. 1982. *Elementary Engineering Fracture Mechanics*. 3rd ed. The Hague, The Netherlands: Martinus Nijhoff Publishers.

Cantwell, W.J. and Morton, J. 1991. The impact resistance of composite materials—A review. *Composites* 22, 347–362.

Carter, P. 1998. Creep failure of a spray drier. *Eng. Failure Anal.* 5, 143–147.

Collins, J.A. 1993. *Failure of Materials in Mechanical Design: Analysis, Prediction, Prevention.* New York: John Wiley & Sons, Inc.

Das, S., Cheng, J.J.R., and Murray, D.W. 2007. Prediction of fracture life of a wrinkled steel pipe subject to low cycle fatigue load. *Canadian J. Civil Eng.* 34, 1131–1139.

Davis, R.J., ed. 2000. *Corrosion: Understanding the Basics.* Materials Park, OH: ASM International, the Materials Information Society.

Degrieck, J. and Paepegem, W.V. 2001. Fatigue damage modeling of fiber-reinforced composite materials: Review. *Appl. Mech. Rev.* 54, 279–300.

Garwood, S.J. 1997. Investigation of the MV Kurdistan casualty. *Eng. Failure Anal.* 4, 3–24.

Henshaw, J.M., Wood, V., and Hall, A.C. 1999. Failure of automobile seat belts caused by polymer degradation. *Eng. Failure Anal.* 6, 13–25.

Hesselman, D.P.H. 1969. Unified theory of thermal shock fracture initiation and crack propagation in brittle ceramics. *J. Am. Ceram. Soc.* 52, 600–604.

Hoeppner, D.W. 1992. Mechanisms of fretting fatigue and their impact on test methods development. In *ASTM STP* 1159, ed. M.H. Attia and R.B. Waterhouse, 23–32, Philadelphia: ASTM.

Hoeppner, D.W. 2006. Fretting fatigue case studies of engineering component. *Tribology Int.* 39, 1271–1276.

Hoeppner, D.W., Chandrasekaran, V., and Elliot III, C.B., eds. 2000. *Fretting Fatigue: Current Technology and Practices. ASTM STP* 1367, West Conshohocken, PA: ASTM.

Kassner, M.E. 2009. *Fundamentals of Creep in Metals and Alloys*, 2nd ed. Amsterdam, The Netherlands: Elsevier.

Kobrick, R.L., Klaus, D.M., and Street, K.W. 2011. Standardization of a volumetric displacement measurement for two-body abrasion scratch test data analysis. *Wear* 270, 650–657.

Koch, G.H., Brongers, M.P.H., Thompson, N.G., Virmani, Y.P., and Payer, J.H. 2002. Corrosion Costs and Preventive Strategies in the United States, Supplement to *Materials Performance*, Report No. FHWA-RD-01-156. Washington, DC: Federal Highway Administration.

Lemaitre, J. and Desmorat, R. 2005. *Engineering Fracture Mechanics: Ductile, Creep, Fatigue and Brittle Fracture.* Berlin Heidelberg: Springer-Verlag.

Massi, F., Rocchi, J., Culla, A., and Berthier, Y. 2010. Coupling system dynamics and contact behaviour: Modelling bearings subjected to environmental induced vibrations and 'false brinelling' degradation. *Mech. Sys. Signal Proc.* 24, 1068–1080.

McCafferty, E. 2010. *Introduction to Corrosion Science.* New York: Springer Science + Business Media, LLC.

National Transportation Safety Board, Aircraft Accident Report, Aloha Airlines, Flight 243, Boeing 737-200, N73711. NTSB Number: AAR-89-03. NTIS Number: PB89-910404, Washington, DC: NTSB.

Naumenko, K. and Altenbach, H. 2007. *Modeling of Creep for Structural Analysis.* Berlin Heidelberg: Springer-Verlag.

Olsson, M. 2011. Tribological evaluation of some potential tribo materials used in column lift rolling contacts—A case study. *Wear* 270, 720–724.

Qi, F., Meng, S., and Guo, H. 2012. Repeated thermal shock behavior of the $ZrB2$-SiC-Zrc ultra-high-temperature ceramic. *Mater. Design* 35, 133–137.

Reed, R.P., Smith, J.H., and Christ, B.W. 1983. Special Publication 647. Gaithersburg, MD: U.S. Department of Commerce, National Bureau of Standards.

Revie, R.W. and Uhlig, H.H. 2008. *Corrosion and Corrosion Control: An Introduction to Corrosion Science and Engineering*, 4th ed. Hoboken, NJ: John Wiley & Sons, Inc.

Richardson, M.O.W. and Wisheart, M.J. 1996. Review of low-velocity impact properties of composite materials. *Composites Part A: Applied Science and Manufacturing* 27, 1123–1131.

Schneider, G.A., ed. 1993. *Thermal Shock and Thermal Fatigue Behavior of Advanced Ceramics.* Dordrecht, The Netherlands: Kluwer Academic Publishers.

Schoen, R.R., Habetler, T.G., Kamran, F., and Bartheld, R.G. 1995. Motor bearing damage detection using stator current monitoring. *IEEE Trans. Industry Appl.* 31, 1274–1279.

Schweitzer, P.A. 2010. *Fundamentals of Corrosion: Mechanisms, Causes and Preventive Methods.* Boca Raton, FL: CRC Press, Taylor & Francis Group.

Stachowiak, G.W., ed. 2005. *Wear–Materials, Mechanisms and Practice.* Chichester, London: John Wiley & Sons Ltd.

Stephens, R.I., Fatemi, A., Stephens, R.R., and Fuchs, H.O. 2000. *Metal Fatigue in Engineering,* 2nd ed. New York: John Wiley & Sons, Inc.

Vogelesang, L.B. and Vlot, A. 2000. Development of fibre metal laminates for advanced aerospace structures. *J. Mater. Proc. Tech.* 103, 1–5.

2 Fundamentals of Thermodynamics

In this chapter, the basic concepts and definitions relevant to thermodynamics of fatigue are presented. Particular attention is given to the notion of entropy generation and its utility in assessment of degradation.

To begin with, it is relevant to point out that in general, degradation processes differ from each other by their *active mechanisms*, which produce irreversible alteration in the material. For example, degradation caused by a fatigue process is essentially due to the creation of dislocations at microscale and generation and propagation of cracks at meso- and macroscale. Here, fatigue mechanism is governed by the application of cyclically repeated mechanical loading or deformation. On the other hand, in a typical sliding wear process, degradation is attributed to the *failure* of the tip of the surface asperities at the contact interface and the associated removal of material. In this case, failure mechanism is due to the presence of mechanochemical forces. However, regardless of the type, the thermomechanical principles describing the behavior of degradation are general and applicable to all degrading processes.

2.1 OPEN AND CLOSED SYSTEMS

The first step in thermodynamic analysis of a problem is to define the region (or matter) of interest. We simply represent this designated region by calling it the *system*; whatever lies outside this *region of interest* is referred to as the *environment* or the *surroundings*. While the system and its environment are separated by an imaginary wall called the system *boundary*, they are intimately connected: energy and matter can exchange between the system and its environment through the system boundary. A system that does not allow transfer of matter is referred to as a *closed system*. A closed system can, however, exchange energy with its surroundings. For example, a metallic sample exposed to fatigue loading as shown in Figure 2.1(a) is basically a closed system, because no mass is exchanged with its environment. In contrast, consider a soft material rubbing against a hard material (tribosystem) shown in Figure 2.1(b). If we consider the soft material as the system of interest, its mass continuously transfers out to the environment as wear fragments are generated and cross the system boundary, making it an *open system*.

Note that in many references, an open system is often referred to as a control volume and a closed system is simply referred to as a system. With our focus on fatigue, where typically there is no mass transfer, we simply refer to the region of interest as the system.

The choice of the system boundary is, to some extent, arbitrary. However, once the boundary is defined, it can significantly influence the treatment of the problem. A comprehensive discussion on the effectiveness of appropriate selection of boundaries in analysis of thermodynamic systems is given by Bejan (1988). To illustrate, consider a specimen (system) in contact with grips (surroundings) as shown in Figure 2.2(a). Let T_S and T_G represent the

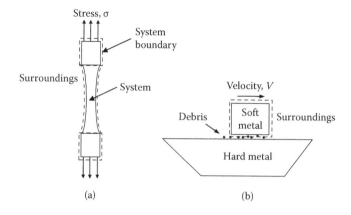

FIGURE 2.1   An example of (a) closed and (b) open systems.

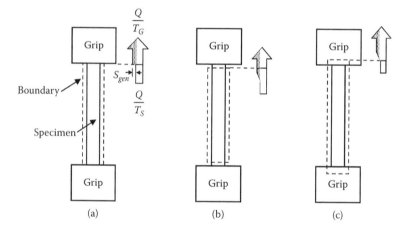

FIGURE 2.2   Heat and entropy transfer through the system's boundary.

temperature of specimen and grip, respectively. Assume $T_G < T_S$, so that heat, Q, is transferred through the boundary from specimen to the grip. As shown in this figure, the amount of heat Q is conserved as it passes through the system boundary. However, entropy transfer,[*] Q/T, is not conserved as it passes through the boundary. The confusion arises because in definition of entropy Q/T, the temperature in the denominator could be either T_G or T_S, and also, the volumetric entropy generation S_{gen} is absent (see Figure 2.2a). The proper choice of the system's boundary as shown in Figures 2.2(b) and 2.2(c) resolves this confusion. In these figures, the boundary is chosen such that temperature changes continuously across it and entropy transfer is conserved. In Figure 2.2(b), the boundary is located inside the specimen (system) and, therefore, the entropy generation due to temperature gradient is considered to be internal. In Figure 2.2(c), the boundary is located outside the specimen and the entropy generation due to temperature gradient is external. The choice of boundary between Figure 2.2(b) and (c) is arbitrary. However, in this book, we consider the entropy generation due to the system–environment interaction to be internal. This means that the temperature gradient within the

---

[*] The concept of entropy S, is discussed in Section 2.6. For this discussion, simply assume $S = Q/T$.

specimen is a source of internal entropy generation. The choice of the system boundary is illustrated schematically in Figures 2.3(a) and (b). Figure 2.3(a) shows a specimen placed within the grips of a fatigue testing machine and Figure 2.3(b) shows the specimen as a system as it is separated from the grips and environment by a system boundary.

2.2 EQUILIBRIUM AND NONEQUILIBRIUM STATE

The *state* of the system, at any instant in time, is identified by its *properties* such as temperature, T, volume, V, and pressure, P. If the instantaneous temperatures at different locations in a system happen to be different (see Figure 2.4), then the state of the system cannot be identified. Here, $T_1 \neq T_2$ and because of the temperature gradient, there exists a *driving force* associated with it so that the *thermal equilibrium* is not maintained. The state of the system and/or the state variables such as entropy, S, internal energy, U, and total energy, E, can only be defined for equilibrium state at which the properties are uniformly distributed. At equilibrium state, all driving forces cease to exist. That is, thermal equilibrium is satisfied when

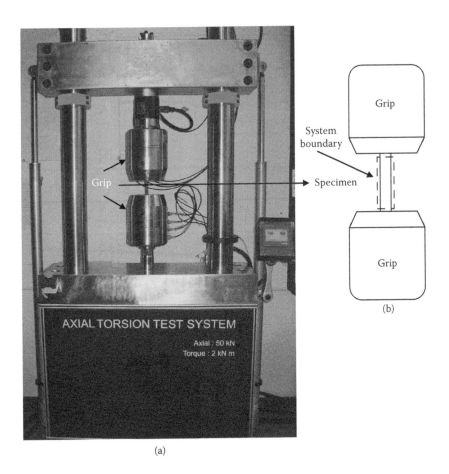

(a)

FIGURE 2.3 (a) Specimen in a fatigue testing machine, (b) as it is considered as a system.

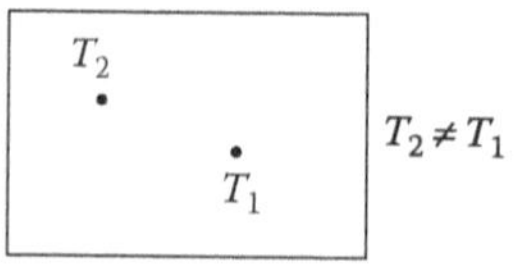

$$T_2 \neq T_1$$

FIGURE 2.4 A closed system in nonequilibrium state.

$T_1 = T_2$ and heat flow vanishes. Similarly, a system is said to be at *mechanical equilibrium* when there is no pressure gradient or at *chemical equilibrium* when there is no chemical potential.

Given that systems in nature are not in equilibrium state, one is compelled to ask: how can the thermodynamics properties such as entropy, internal energy, and so forth, be described? To overcome this difficulty, the concept of *local equilibrium* is invoked wherein the system is assumed to be divided into subsystems (or cells) whose temperature, stress, and entropy behave as if the system operates in equilibrium in a local sense (Yourgrau, Van Der Merwe, and Raw 1982). In this sense, each cell can be treated as an equilibrium system even if the entire system is not in equilibrium state. According to Jou, Casas-Vazquez, and Lebon (2010), the notion of the local equilibrium assumes that

(a) All the variables are defined just as they are in equilibrium and that the properties of each cell remain uniform but different from cell to cell; and
(b) The equilibrium relationships between entropy and *state variables* remain valid outside equilibrium.

The notion of local equilibrium, thus, provides a powerful tool for analyzing nonequilibrium systems. In Chapters 5 and 6, we employ this concept to assess a fatigue system.

2.3 STEADY AND UNSTEADY STATE

Steady-state (or stationary) condition is attained when the state variables remain constant as time passes. A system whose state variables change with time is said to be an *unsteady* system. The difference between steady-state and equilibrium state is that in steady-state one deals with uniformity in time coordinate, while equilibrium implies uniformity in space coordinate. Consider A to be a state property (such as entropy). If t represents time and x denotes space, then

$$\text{at steady-state condition:} \frac{dA}{dt} = 0$$

$$\text{at equilibrium condition:} \frac{dA}{dx} = 0$$

An equilibrium system can be either steady or unsteady and similarly, a steady system can be either in an equilibrium or nonequilibrium state. Note that for steady-state condition to hold, the state variables must be invariant with time; variables that are not state variables (e.g., heat and work) may or may not change with time. These *non-state* variables are known as path dependent variables and will be studied in Section 2.5. There, it is shown that the first law of thermodynamics, written as a function of time, can be expressed as

$$\frac{dU}{dt} = \frac{\delta Q}{dt} - \frac{\delta W}{dt}$$

where U is the internal energy, Q and W are heat and work exchange with surroundings, respectively. At the steady-state condition, the rate of internal energy vanishes, $dU/dt = 0$; however, both $\delta Q/dt$ and $\delta W/dt$ can be nonzero.

Steady-state is an important condition with some special traits. For example, as discussed in Section 2.7, at steady-state, the rate of change of entropy vanishes resulting in a balance between entropy generation and entropy flow. This constraint on the system provides a powerful tool to evaluate entropy generation from entropy flow. The entropy flow can be conveniently evaluated from experimental measurements, while entropy generation is linked to internal variables that may not be easily measurable.

2.4   STABLE AND UNSTABLE STATE

Stability is another important topic in thermodynamic analysis, for it can be used to define the trajectories of the evolution of a system in time. In simple words, a system is said to be stable if it reacts to the environmental perturbation in a way that reverts back to the undisturbed condition. For example, a pendulum at its lowest position is stable because if it is perturbed either to the left or right, it reverts back to its original position. The stability criterion is also connected to the *Le Chatelier's Principle* which states that "any inhomogeneity that somehow develops in the system should induce a process that tends to eradicate the inhomogeneity" (Callen 1985). In essence, the system tends to resist the effect of environmental disturbance.

The stability analysis based on the entropy generation is the central principle upon which the notion of self-organization and the physics of highly ordered systems are built. The stability of nonequilibrium stationary state can be determined based on the *Lyapunov's* theory of stability (cf. Chapter 8).

In what follows, the first and the second laws of thermodynamics are discussed, and fundamental entropy formulations required for the analysis of fatigue process are derived.

2.5   FIRST LAW OF THERMODYNAMICS

The first law of thermodynamics states that total energy of a system is conserved, even though energy can transform from one form to another. For example, the mechanical energy input to a system can convert into heat and/or stored energy in the material. The first law of thermodynamics simply enforces a balance between the total energy E, work W, and heat Q:

$$dE = \delta Q - \delta W \tag{2.1}$$

Referring to Figure 2.5, the sign convention is adopted where work is considered to be positive if it is done by the system and heat is positive if it enters into the system. Heat and work are both inexact quantities, meaning that they are both path dependent, and therefore, they

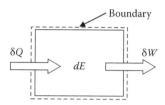

FIGURE 2.5   Balance of energy for a closed system.

cannot be expressed by specifying the initial and final states of the system alone. Hence, we show their differentials in the inexact form δQ and δW. In contrast, the total energy, E, is a system property and depends only on the initial and final conditions. Therefore, its differential is written in the exact form dE.

The system's energy E consists of kinetic energy KE, potential energy PE, and the internal energy U. Therefore, Equation (2.1) can be written as

$$dU + d(PE) + d(KE) = \delta Q - \delta W \tag{2.2}$$

In fatigue analyses, typically the kinetic and potential energy are neglected, that is, $KE = PE = 0$. Therefore, Equation (2.2) is rewritten as

$$dU = \delta Q - \delta W \tag{2.3}$$

The internal energy U is also a state variable and depends only on the initial and final conditions. Equation (2.3) can be expressed in volumetric (or specific) form by dividing both sides of the equation by the mass of the system m, as

$$du = \delta q - \delta w \tag{2.4}$$

where $u = U/m$, $q = Q/m$, and $w = W/m$.

The first law of thermodynamics can also be written in time rate form. If the incremental change in the system's internal energy du takes place during the time interval δt, the rate equation form of the first law can be written as

$$\frac{du}{dt} = \dot{q} - \dot{w} \tag{2.5}$$

where $\dot{q}$ is the rate of heat transfer and $\dot{w}$ is the power transfer. At the steady-state condition during which the internal energy remains constant with time, that is, $du/dt = 0$, Equation (2.5) reduces to

$$\dot{q} = \dot{w} \tag{2.6}$$

For solid materials, the change in internal energy is related to the change in temperature as

$$dU = mc\,dT \tag{2.7}$$

where c is the specific heat capacity. Note that, in general, for solids the constant-pressure and the constant-volume specific heats are nearly the same, that is, $c_p \approx c_v = c$.

Having the temperature evolution of a solid material during a process, the evolution of its internal energy can be evaluated from Equation (2.7). The following example shows implementation of the first law for a fatigue process. The data given in the example are adopted from experimental work of Jiang et al. (2001).

Example 2.1

A solid, cylindrical-shape metal specimen initially at room temperature, $T_0 = 296$ K, is subjected to cyclic mechanical loading at frequency of $f = 20$ Hz. The dimensions of the specimen are given in Figure 2.6. The density and specific heat capacity of the metal are $\rho = 8470$ kg/m³ and $c = 456$ J/kg.K, respectively. The density of mechanical input work per cycle is 75 kJ/m³.per cycle. Experiment shows that the resulting temperature rise can be described by $T(t) = 296 + 17$ $(1 - e^{-0.0194\,t})$, where t is time. For simplicity, assume that the entire gage length of the specimen undergoes the same temperature evolution, that is, no temperature gradient in the specimen.

(a) Plot the temperature evolution as a function of the number of cycle N, from $N = 0$ to $N = 97{,}500$. Find the number of cycles at which the temperature reaches a steady-state condition.
(b) Use an appropriate form of the first law to find the evolution of heat dissipation Q, and plot the results from $N = 0$ to $N = 15{,}000$.
(c) If the convection heat transfer coefficient h, is given to be $h = 12.5$ W/m²K, find the amount of heat transferred to the surroundings by means of convection from $N = 15{,}000$ to $N = 35{,}000$. Estimate the contribution of conduction and radiation heat transfer.

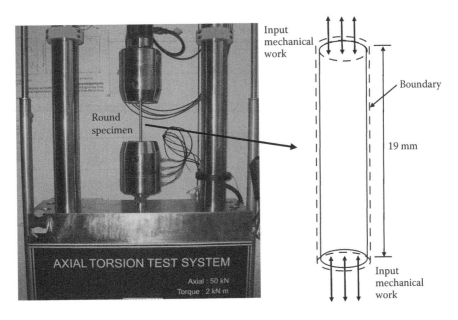

FIGURE 2.6 Geometry of the specimen (Diameter, $D = 5$ mm).

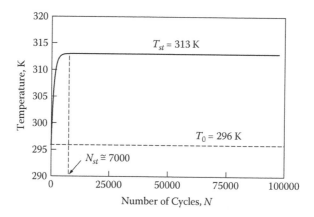

FIGURE 2.7   Temperature evolution as a function of number of cycles.

SOLUTION

Let us first consider the entire gage length as the system boundary as shown in Figure 2.6.

(a) Solution is very simple; we just need to plot the temperature as a function of number of cycles by replacing time in $T(t) = 296 + 17\,(1 - e^{-0.0194t})$ with $t = N/f$, where f is frequency. Therefore, we have $T(N) = 296 + 17$ $(1 - e^{-0.00097\,N})$. Figure 2.7 shows the evolution of temperature from $N = 0$ to $N = 97{,}500$. This figure shows that the temperature of the specimen reaches the stationary condition $T_{st} = 313\mathrm{K}$ after almost $N_{st} = 7{,}000$ cycles. The number of cycles at which the specimen fails is given by Jiang et al. (2001) to be $N_f = 97{,}500$. Therefore, the transient phase of temperature rise is less than 10% of the total fatigue life. This subject is discussed in detail in Chapter 4.

(b) Since the input work is given in rate form, the proper form of the first law would be the rate equation form given by Equation (2.5) as

$$\dot{U} = \dot{Q} - \dot{W} \tag{2.8}$$

Therefore, the rate of heat exchange with surroundings can be written as

$$\dot{Q} = \dot{U} + \dot{W} \tag{2.9}$$

Using the expression for the internal energy from Equation (2.7):

$$\dot{Q} = mc\frac{dT}{dt} + \dot{W} \tag{2.10}$$

Substituting the values given in the problem, Equation (2.10) yields

$$\dot{Q} = 8470 \times \frac{\pi}{4} \times 5^2 \times 19 \times 10^{-9} \times 456 \times 0.3298 \times e^{-0.0194t} \tag{2.11}$$

$$-75 \times 20 \times \frac{\pi}{4} \times 5^2 \times 19 \times 10^{-9} = 0.475 \times e^{-0.0194t} - 0.559 \text{ W}$$

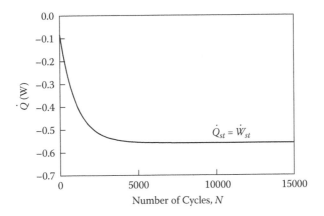

FIGURE 2.8 Evolution of heat exchange with surroundings as a function of number of cycles.

Equation (2.11) determines the evolution of the heat exchange with the surroundings. Figure 2.8 shows the variation of $\dot{Q}$ from $N = 0$ to $N = 15,000$ cycles. Note that in Equation (2.11) the rate of input mechanical work is negative because work is done on the system. Also, the rate of heat exchange is negative since heat it transferred to the surroundings from the system. Therefore, the absolute value of the heat exchange increases as the number of cycles passes until it reaches the value of input mechanical work. Physically, as work is done on the system, both the internal energy and heat exchange increase until the system attains steady-state and the internal energy becomes constant.

The rate of heat transfer, $\dot{Q}$, is shown in Figure 2.8. It includes three modes of heat transfer, convection and radiation heat transfer to the environment and conduction from the specimen to the grips.

(c) Since steady-state is reached during $N = 15,000$ to $N = 35,000$ cycles, the temperature remains constant T_{st}. Therefore, the rate of convection heat transfer to the surroundings can be evaluated from

$$\dot{Q}_{conv} = hA(T_{st} - T_0) \tag{2.12}$$

where h is the convection heat transfer coefficient and A is the surface area. Substituting the given values into Equation (2.12), we have

$$\dot{Q}_{conv} = 12.5 \times \pi \times 5 \times 19 \times 10^{-6} \times (313 - 296) = 0.063 \text{ W} \tag{2.13}$$

As mentioned earlier, the total heat dissipation includes conduction, convection, and radiation heat transfer modes, that is,

$$\dot{Q}_{st} = \dot{Q}_{cond} + \dot{Q}_{rad} + \dot{Q}_{con} \tag{2.14}$$

Having determined $\dot{Q}_{conv}$ from Equation (2.13) and knowing that at stationary state, the rate of heat dissipation balances the rate of input mechanical work, $\dot{Q}_{st} = \dot{W}_{st}$, Equation (2.14) gives

$$\dot{Q}_{cond} + \dot{Q}_{rad} = 0.559 - 0.063 = 0.496 \text{ W} \qquad (2.15)$$

The accumulation of convection heat transfer during $N = 15{,}000$ to $N = 35{,}000$ cycles can now be evaluated:

$$Q_{conv} = 0.063 \times \frac{(35000 - 15000)}{20} = 63 \text{ J} \quad \blacktriangle$$

In the preceding example, we applied the first law of thermodynamics to a closed system in which mass transfer is not allowed. However, in a general sense, one may have to deal with open systems, as shown in Figure 2.9, with mass flow inlet and outlet. Unlike a closed system where the quantity of mass remains constant within the system, in an open system, mass can flow into the system at the rate of $\dot{m}_{in}$ and exit the system at $\dot{m}_{out}$. In general, of course, there may be several sources entering and leaving. The mass influx can also bring in energy and work to the system or can take them out of the system. The amount of internal energy and work associated with flow of mass is expressed by the total enthalpy of the flow, h_{tot}. Further discussion on the interpretation of total enthalpy is beyond the scope of this book. Interested readers may refer to Bejan (1987).

Considering the mass flow in and out of the system, the first law of thermodynamics in the rate form for the open system shown in Figure 2.9 can be written as

$$\frac{dE}{dt} = \dot{Q} - \dot{W} + (\dot{m}h_{tot})_{in} - (\dot{m}h_{tot})_{out} \qquad (2.16)$$

For a closed system $\dot{m}_{in} = \dot{m}_{out} = 0$ and Equation (2.5) reduces to

$$\frac{dE}{dt} = \dot{Q} - \dot{W}$$

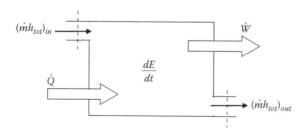

FIGURE 2.9   Mass flow inlet and outlet for an open system.

2.6 SECOND LAW OF THERMODYNAMICS

Energy conversion is always accompanied by irreversibility, making it impossible for the system to return back to its initial state. In contrast, an idealized reversible process after a time interval Δt can revert back to its initial state with time replaced by $-\Delta t$. The following succinct definition is proposed by Sonntag, Borgnakke, and Van Wylen (2003), "A reversible process for a system is defined as a process that once having taken place can be reversed and in so doing leave no change in either system or surroundings." This is, of course, purely an idealization. To illustrate, consider one complete loading cycle in a tension–compression test corresponding to Example 2.1. Stress is applied on the system (specimen) in the first half-cycle and as a result the mechanical work converts into heat, increasing the temperature of the specimen. Let us assume that hypothetically the cycle is performed so slowly that the temperature reduces to initial state after the second half-cycle. That is, the specimen returns back to its initial shape. During this complete cycle, even though the specimen is restored into its initial state, the state of the surroundings has changed since heat is transferred to the surroundings. The heat transferred to the surroundings cannot be restored and converted to useful mechanical work. Therefore, practically the process of cyclic loading the specimen is an irreversible one.

In the foregoing example about cyclic loading of a specimen, the direction of the process is such that, unless facilitated by an external element, the dissipated heat (low-quality energy) cannot convert into the available mechanical work (high-quality energy). The possibility and direction of a process is the main subject of the second law of thermodynamics. For this purpose, a quantity called entropy—popularly known as the *arrow of time*—can be invoked to determine the evolution of the system during an irreversible process.

Nearly all processes in nature are irreversible, and this irreversibility gives rise to disorder. In an isolated system, the level or the state of disorder always increases until attaining its maximum value once equilibrium is reached. To illustrate, consider an *organized* set of molecules enclosed by an imaginary box in an isolated room as shown in Figure 2.10. At time $t = t_1$, we open the box and allow it to communicate with surroundings such that molecules begin to move out of the box and occupy the room. Clearly, this process is irreversible. In other words, at time $t_2 > t_1$, molecules are randomly distributed in the room resulting

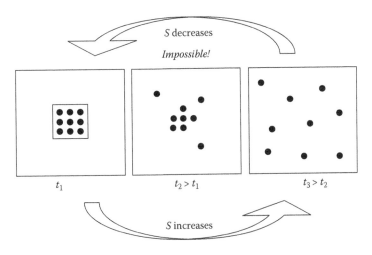

FIGURE 2.10 Increase in entropy for an isolated system.

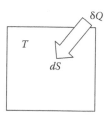

FIGURE 2.11   Change in the entropy of the system.

in an increase in disorder, and practically there is no possibility that all molecules move back to the initial organized state if time t is replaced by $-t$. Letting P denote the number of ways in which molecules can be distributed in the room, one can quantify the associated entropy using the following equation:

$$S = k \ln P \qquad (2.17)$$

where k is the Boltzmann's universal constant. The second law of thermodynamics (Boltzmann 1896) asserts that the entropy of the system, which embodies the isolated room and molecules, will increase as time passes until equilibrium is attained at $t = t_3$ and the molecules are uniformly distributed. This simple example illustrates utility of the concept of entropy as a powerful probe for the study of irreversible processes.

2.7   ENTROPY FLOW AND ENTROPY GENERATION

In classical thermodynamics, one can associate to any macroscopic system a state function S that represents the entropy. Consider the system shown in Figure 2.11, which receives heat δQ from surroundings. According to *Clausius Inequality,*[*] the change in a system's entropy, dS, as it undergoes a change of state is related to δQ by

$$dS \geq \frac{\delta Q}{T} \qquad (2.18)$$

where T is the temperature in Kelvin.

The equality in Equation (2.18) holds only for the reversible processes. The reversible entropy $\delta Q/T$ is also referred to *entropy flow* d_eS. To account for irreversible processes, we convert the inequality to a balance equation by adding an extra term to the right-hand side of Equation (2.18) and removing the inequality sign for quality:

$$dS = \frac{\delta Q}{T} + d_iS \qquad (2.19)$$

---

[*] The second law of thermodynamics was first discovered by R. Clausius in 1865 and developed by Boltzmann in 1896.

The extra term d_iS represents the internal irreversibility in the system.[*] Substituting $\delta Q/T = d_eS$ in Equation (2.19), we have

$$dS = d_eS + d_iS \qquad (2.20)$$

where d_eS is the *entropy exchange* (flow) with surroundings and d_iS is the *entropy generation* inside the system. Note that unlike the entropy S, the entropy flow d_eS, and entropy generation d_iS are not state functions, as they are path dependent. Therefore, differentials on the right-hand side of Equation (2.20) are inexact. Comparing Equation (2.20) with Equation (2.18) yields

$$d_iS \ge 0 \qquad (2.21)$$

which is known as the second law of thermodynamics.

While d_iS is always positive, d_eS may be negative or positive, and if the system is isolated, it is necessarily zero. If d_eS is *pumped* to the surroundings, that is, $d_eS < 0$ and the following condition is satisfied

$$|d_eS| > d_iS \ge 0 \qquad (2.22)$$

then the overall entropy of the system decreases, that is, $dS < 0$. This implies that via the influence of an external element(s) as the driving force(s), it is possible to export entropy to the surroundings. The decrease in entropy can result in the establishment of order and the formation of less random structures. Regardless of whether entropy is exported to the surroundings, the entropy generation within the system, d_iS, always increases, and it is this entropy that, in a sense, characterizes exhaustion of the system's lifespan. In other words, the *degradation* or *aging* of the system is intimately connected to the entropy production that accumulates progressively over time and degrades the system until its final breakdown (Bryant, Khonsari, and Ling 2008; Naderi, Amiri, and Khonsari 2010). This subject will be discussed in detail in Chapters 5 and 6.

Equation (2.20) can be expressed in the rate form by dividing both sides by dt:

$$\frac{dS}{dt} = \frac{d_eS}{dt} + \frac{d_iS}{dt} \qquad (2.23)$$

If a system is in steady-state condition, the entropy does not depend on time, and from Equations (2.20) and (2.21), it follows that

$$d_eS + d_iS = 0; \quad d_eS = -d_iS < 0 \qquad (2.24)$$

Equation (2.24) implies that at the steady-state condition the entropy flow leaving the system is equal to the entropy generated within the system. For experimental purposes, entropy flow is often easier to measure compared to entropy generation. In the next example, we illustrate how the entropy flow evolves in a system that undergoes cyclic mechanical loading.

---

[*] The notations d_iS and d_eS are employed to emphasize that these inexact quantities are different from total (material) derivatives dS_i and dS_e.

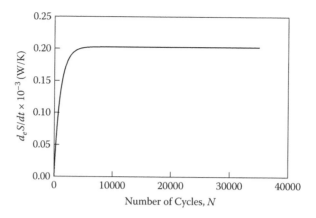

FIGURE 2.12 Evolution of the rate of entropy flow.

Example 2.2

Referring back to the specimen of Example 2.1, calculate the evolution of entropy flow to the environment, d^eS/T, from $N = 0$ to $N = 35{,}000$ cycles. Also, evaluate the evolution of accumulated entropy flow from $N = 0$ to $N = 35{,}000$ cycles.

SOLUTION

Having determined the rate of convection heat transfer and evolution of temperature, we can evaluate the rate of entropy flow as

$$\frac{d_eS}{dt} = \frac{1}{T}\frac{dQ_{conv}}{dt} = \frac{hA(T - T_0)}{T} \tag{2.25}$$

The convection heat transfer coefficient h is given to be $h = 12.5 \ \text{W/m}^2\text{K}$, and temperature is $T(t) = 296 + 17\,(1 - e^{-0.0194\,t})$ with $T_0 = 296$ K. Figure 2.12 shows the evolution of entropy flow from $N = 0$ to $N = 35{,}000$. It shows that the rate of entropy flow has a similar trend as temperature (cf. Figure 2.7). At steady-state, the rate of entropy flow is equal to the rate of entropy generation. However, the accumulation of entropy flow which is the integration of Equation (2.25) is linearly increasing by the number of cycles, as shown in Figure 2.13. The linear trend of accumulation of entropy flow provides a powerful technique for evaluation of damage in fatigue problems. This concept will be discussed in Chapter 5. ▲

2.8 ENTROPY BALANCE EQUATION

In Example 2.2, we demonstrated that entropy flow can be easily measured experimentally. In contrast, measurement of entropy generation is not that simple. Therefore, it is highly desirable to come up with a suitable relationship that can be directly related to measurable quantities inside the system.

Consider a system of solid medium that undergoes two irreversible processes involving heat conduction and deformation. These two processes are the major sources of irreversibility

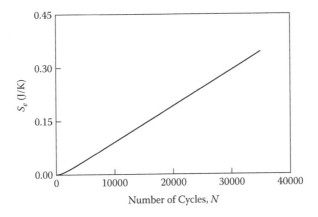

FIGURE 2.13 Evolution of the of entropy flow.

in fatigue problems that will be covered in Chapter 3. Based on the local equilibrium concept, we can divide the system into subsystems (cells), each having uniform properties. For simplicity of analysis, let us assume that the system has only two subsystems, 1 and 2, each maintained at uniform temperature T_1 and T_2 with $T_2 > T_1$ (see Figure 2.14). Owing to the temperature difference, heat $\delta Q'$ is transferred from subsystem 2 to 1. We assume that external work deforms each subsystem and that as a result of deformation, heat is generated within each subsystem. Let δQ_{def} represent the heat generation inside the subsystem by an external heat supply. This way we remove irreversibility inside the system by introducing an external heat sources.

Since entropy is an extensive property, the entropy of the whole system is the summation of entropy of both subsystems:

$$dS = (dS)_1 + (dS)_2 \qquad (2.26)$$

We now write the net heat received by each subsystem as

$$(dQ)_1 = -(dQ_{ex})_1 + (dQ_{def})_1 + dQ', \quad (dQ)_2 = -(dQ_{ex})_2 + (dQ_{def})_1 - dQ' \qquad (2.27)$$

where dQ_{ex} is the heat exchange with surroundings, dQ_{def} is the heat supply due to external work, and dQ' is the heat transfer between two subsystems due to temperature difference.

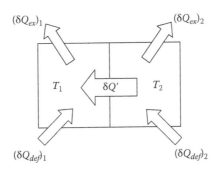

FIGURE 2.14 Entropy generation in a system with heat conduction and deformation.

Taking into account Equation (2.19) for reversible systems, the entropy change of the whole system becomes

$$dS = \frac{(dQ)_1}{T_1} + \frac{(dQ)_2}{T_2} \tag{2.28}$$

$$dS = -\frac{(dQ_{ex})_1}{T_1} - \frac{(dQ_{exc})_2}{T_2} + dQ'\left(\frac{1}{T_1} - \frac{1}{T_2}\right) + \frac{(dQ_{def})_1}{T_1} + \frac{(dQ_{def})_2}{T_2} \tag{2.29}$$

In agreement with Equation (2.19), the entropy change consists of two parts. The first:

$$d_eS = -\frac{(dQ_{ex})_1}{T_1} - \frac{(dQ_{exc})_2}{T_2} \tag{2.30}$$

represents the exchange of heat with surroundings, while the second part:

$$d_iS = dQ'\left(\frac{1}{T_1} - \frac{1}{T_2}\right) + \frac{(dQ_{def})_1}{T_1} + \frac{(dQ_{def})_2}{T_2} \tag{2.31}$$

results from the irreversible heat conduction inside the system and deformation due to mechanical work. It is interesting to mention that the entropy generation due to heat conduction is driven by the gradient of inverse temperature. The entropy production due to conduction can be zero only after thermal equilibrium is established, that is, $T_1 = T_2$. The detailed derivation of entropy generation due to different irreversible processes will be given in Chapter 3.

We now present the entropy balance equation in general form for irreversible processes that occur in continuous systems containing spatial nonuniformities. Under the hypothesis of local equilibrium, the entropy balance equation can be expressed in local form as follows:

$$\rho\frac{ds}{dt} = \sigma - \nabla \cdot \boldsymbol{J}_{s,tot} \tag{2.32}$$

where

$$S = \int_V \rho s \, dV \tag{2.32a}$$

$$\frac{d_eS}{dt} = -\int_\Omega \boldsymbol{J}_{s,tot} \, d\Omega \tag{2.32b}$$

$$\frac{d_iS}{dt} = \int_V \sigma dV \tag{2.32c}$$

where s is the entropy per unit mass, ρ is the density, $\boldsymbol{J}_{s,tot}$ is the total entropy flow per unit area and unit time, $d\Omega$ is the element of surface area, and σ denotes the *entropy production per unit volume per unit time*. The second law of thermodynamics requires

$$\sigma \geq 0 \tag{2.33}$$

where the equality in Equation (2.33) holds only when the process is reversible. Let us assume that the gradient of chemical potential within the system is negligible so that the total entropy flow per unit area and per unit time simply reduces to the heat flux as

$$J_{s,tot} = \frac{J_q}{T} \qquad (2.34)$$

Upon the substitution of Equation (2.34) into Equation (2.32), we obtain

$$\rho \frac{ds}{dt} = - \cdot \left(\frac{J_q}{T} \right) + \sigma \qquad (2.35)$$

The second term on the right-hand side of Equation (2.35) can be written as

$$\cdot \left(\frac{J_q}{T} \right) = \frac{\cdot J_q}{T} - J_q \cdot \frac{T}{T^2} \qquad (2.36)$$

Finally substituting Equation (2.36) into Equation (2.35) yields

$$\rho \frac{ds}{dt} = - \frac{1}{T} \cdot J_q + J_q \cdot \frac{T}{T^2} + \sigma \qquad (2.37)$$

Equation (2.37) is the balance equation for volumetric entropy ρs with the entropy generation source σ, which satisfies Equation (2.33). In Chapters 5 and 6, we will use Equation (2.37) to develop the constitutive equation of system undergoing cyclic deformation.

REFERENCES

Bejan, A. 1988. *Advanced Engineering Thermodynamics*. New York: John Wiley & Sons, Inc.

Boltzmann, L. 1896. *Vorlesungen uber Gastheorie*. Leipzig: J.A. Barth.

Bryant, M.D., Khonsari, M.M., and Ling, F.F. 2008. On the thermodynamics of degradation. *Proc. R. Soc. A* 464, 2001–2014.

Callen, H.B. 1985. *Thermodynamics and an Introduction to Thermostatistics*. New York: John Wiley & Sons, Inc.

Jiang, L., Wang, H., Liaw, P.K., Brooks, C.R., and Klarstrom, D.L. 2001. Characterization of the temperature evolution during high-cycle fatigue of the ULTIMET superalloy: Experiment and theoretical modeling. *Metall. Mater. Trans. A* 32, 2279–2296.

Jou, D., Casas-Vazquez, J., and Lebon, G. 2010. *Extended Irreversible Thermodynamics*, 4th ed. Berlin and Heidelberg: Springer-Verlag GmbH & Co.

Naderi, M., Amiri, M., and Khonsari, M.M. 2010. On the thermodynamic entropy of fatigue fracture. *Proc. R. Soc. A* 466, 423–438.

Sonntag, R.E., Borgnakke, C., and Van Wylen, G.J. 2003. *Fundamentals of Thermodynamics*. New York: John Wiley & Sons, Inc.

Yourgrau, W., Van Der Merwe, A., and Raw, G. 1982. *Treatise on Irreversible and Statistical Thermophysics: An Introduction to Nonclassical Thermodynamics*. New York: Dover Publications.

3 Degradation–Entropy Generation (DEG) Theorem

In this chapter, we apply the thermodynamics concepts developed in Chapter 2 and present a degradation–entropy generation relationship with particular interest to fatigue. Also introduced in this chapter are the concepts of thermodynamic forces and fluxes. Definitions of new concepts, and degradation forces and fluxes are given, and their affiliation with thermodynamic counterparts is discussed. We proceed to formulate degradation processes within a thermodynamic framework where entropy generation is treated as a measure of exhaustion of the system's lifespan. In other words, the *degradation* or *aging* of the system is intimately connected to the entropy production that accumulates progressively over time and degrades the system until its final breakdown.

All types of permanent degradation or aging are irreversible processes that disorder a system and generate irreversible entropy, in accordance with the second law of thermodynamics. Therefore, quite naturally, entropy can be used in a fundamental way to quantify the behavior of irreversible degradation processes. For example, in a fatigue problem, irreversible entropy is generated by dissipation associated with plastic deformation. Similarly, other forms of degradation such as wear (Klamecki, 1980a, 1980b, 1982, 1984; Zmitrowicz 1987a, 1987b, 1987c), and fretting (Dai and Xue 2009) are consequences of irreversible processes that tend to add disorder to the system.

In all dissipative processes, the *Helmholtz free energy*, θ, is responsible for doing useful work. The free energy decays with time. Hence, if θ_i denotes the free energy of a system prior to exposure to a dissipative process, then after the process, its free energy drops to θ_f such that $\theta_f < \theta_i$. This fact is schematically shown in Figure 3.1. The system's degradation continues until it reaches a minimum at the equilibrium state in accordance with what is known as the principle of minimum free energy. Now, given that the free energy and the internal energy of a system are related via temperature and entropy by $\theta = u - Ts$, it immediately follows that the system's path to minimum free energy is always accompanied by increasing entropy until it reaches its peak value at the equilibrium state.

The evolution of dissipative processes can be described using the principles of irreversible thermodynamics, which are, in fact, established based on the concept of entropy generation. It is, therefore, not surprising that the relationship between degradation and entropy generation has been of interest in different research communities (e.g., fatigue and fracture, tribology, materials science, biology, and chemistry) for decades. This recognition has recently led to creation of an experimentally verified Thermodynamic Degradation Paradigm (Bryant, Khonsari, and Ling 2008), where a relationship was established between degradation and entropy generation produced by accompanying irreversible processes occurring in a system. The driving impetus for the development of the Degradation–Entropy Generation (DEG) theorem is the fundamental realization that entropy monotonically increases and free energy monotonically decreases for *all* natural processes and that an entropy-generating irreversible process accompanies all aging phenomena.

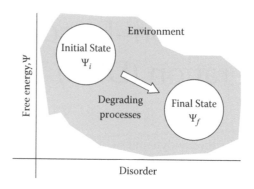

FIGURE 3.1 Degradation processes increase the entropy and reduce the available useful work.

3.1 THERMODYNAMIC FORCES AND FLOWS

An irreversible process is accompanied by transformation of energy, and this transformation is governed by the laws of thermodynamics. Simply put, the first and the second laws of thermodynamics describe the *quantity* and the *quality* of the energy transformation, respectively. The first law or the law of conservation of the energy asserts an equality (balance) between different types of energy (e.g., potential, kinetic, or heat) transferred in a system and quantifies the amount of energy transformation. The second law states that the available work dissipates to "lower-quality energy" in the form of heat. It follows, therefore, that in the course of an irreversible process, the energy of high quality degrades to the low-quality energy. The second law dictates a hierarchy among the different types of energy. One can intuitively imagine that, unless facilitated by an external element, the dissipated heat (low-quality energy) during the fatigue process cannot convert into available mechanical work (high-quality energy). The irreversibilities associated with the energy conversion are represented and discussed in terms of *dissipation* as a measure of irreversibility.

A reversible process can be completely characterized by the specific free energy function ψ. For example, the specific free energy of a solid subjected to elastic deformation can be determined as

$$\psi = \psi(\varepsilon_{ij}, T) \tag{3.1}$$

where ε_{ij} is the component of strain tensor and T is the temperature. We show how the constitutive equations governing the complete thermomechanical behavior of the elastic solid can be derived from Equation (3.1). However, consideration of the specific free energy alone cannot be used to derive the equations that govern a material's behavior. That is because the thermodynamic potentials (e.g., Helmholtz free energy and *Gibbs free energy*[*]) are state functions; they are independent of path and are defined only at equilibrium states. State functions are basically identified and described by a set of independent state variables such as temperature, T, volume, V, and pressure, P. They are sometimes referred to as observable primitive properties since they are measurable quantities (Annamalai and Puri 2002).

In general, irreversible processes possess some common characteristics such as presence of heat conduction, presence of dissipative stresses (Ziegler 1983), and presence of internal

---

[*] The Gibbs free energy is defined as $G = H-TS$, where H denotes the enthalpy.

variables (Lebon, Jou, and Casas-Vazquez 2008). In contrast to observable variables, the internal variables—sometimes referred to as the hidden variables—are not controllable and their evolution is intimately connected to the dissipative processes that take place in the system. While state variables such as temperature, pressure, and volume are macroscopically defined properties, the internal variables are representative of the microscopic features associated with the internal structure. For example, in dealing with plastic deformation of solids, the density of dislocations is identified as an internal variable. Other examples of internal variables can be found in Lebon, Jou, and Casas-Vazquez.

From the foregoing discussion, it can be concluded that to characterize irreversible processes, the specific free energy should be supported with a complementary function which takes into account the effect of, for example, internal variable, plastic deformation, and heat conduction. The complementary function is represented by the *dissipation potential*, Φ, as a measure of exhaustion of the system's lifespan. Note that the dissipation function is not a state function. Unlike the free energy function, which describes only the initial and final states of the system, Φ describes the evolution of the system during the process. Therefore, for a given system, Φ depends on the process path.

Consider, for example, a solid block sliding a distance Δx on a surface with velocity, V, as shown in Figure 3.2. The dissipative process in sliding is friction and the amount of work during the process is

$$\Delta W = -F_f \Delta x \qquad (3.2)$$

where the negative sign means the direction of the friction force is opposite to that of the sliding. The rate of energy dissipation during sliding, $\Delta W/\Delta t$, is

$$\Phi = \frac{W}{t} = -F_f \frac{x}{t} = \vec{F}_f \cdot \vec{V} \qquad (3.3)$$

It can be seen from Equation (3.3) that the dissipation function during this sliding friction process is defined as the inner product of two vectors: the nonconservative friction force, F_f, and the rate of a time-dependent kinematical parameter, $\Delta x/\Delta t$. The product defines the dissipation of energy.

This example shows that, in general, it is possible to define the dissipation function as the product of the dissipative forces and the rate of kinematic parameters. Using thermodynamic terminology, dissipative forces are represented by the generalized thermodynamic forces, X_i^d, and the rate of kinematical parameters by the generalized thermodynamic fluxes, J_i, $i = 1, 2, 3, \ldots, n$. The superscript d stands for dissipative and the index i refers to number of different processes acting in the system. Thermodynamic forces are the driving

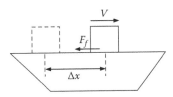

FIGURE 3.2 Energy dissipation during the friction process.

potentials for thermodynamic flows. That is, thermodynamic forces are associated with the gradients of the intensive variables (e.g., temperature and pressure) and cause the *flow* of some quantities within the system.

The product of a pair of forces and fluxes is a measure of energy dissipation and production of entropy. Given that the dissipation function is a function of the thermodynamic fluxes J_i, that is, $\Phi = \Phi(J_i)$, it follows that thermodynamic forces are functions of thermodynamic fluxes:

$$X_i^d = X_i^d(J_j) \quad (i, j = 1, 2, 3, \ldots n)$$

(3.4)

Note, however, that in general, thermodynamic forces depend not only on their conjugate fluxes, but also on all other fluxes. This is discussed in Section 3.2.

The general form of the dissipation function is

$$\Phi = X_i^d(J_j)J_i$$

(3.5)

where the Einstein summation notation is used.

If there are only two dissipative processes, the dissipation function takes on the following form:

$$\Phi = X_1^d J_1 + X_2^d J_2$$

(3.6)

Of course, thermodynamic forces and flows can be scalars, vectors, or tensors. So, the product of $X_i^d(J_j) J_i$ in Equation (3.5) stands for the scalar product of two scalars, the inner product of two vectors, or the double product of two tensors.

The dissipation function, Φ, is related to the entropy generation via $\Phi = T\sigma$, where T is temperature and σ is the volumetric entropy generation. Substituting $T\sigma$ into Equation (3.5) yields

$$\sigma = \frac{X_i^d(J_j)}{T} J_i = X_i(J_j)J_i$$

(3.7)

where we substituted X_i^d/T by X_i and dropped the superscript d for brevity.

It is to be noted that the definition of thermodynamic forces and fluxes is to some extent arbitrary. For instance, in Equation (3.7) one can include the factor $1/T$ in the flux instead of the force (Jou, Casas-Vazquez, and Lebon 2010). However, following the conventional notation in the literature, we consider the factor $1/T$ in the forces instead of the flows, without limiting the generality.

In the following example, we illustrate how the thermodynamic force, X, and flux, J, can be obtained in a heat conduction process.

Example 3.1

Consider the steady-state, one-dimensional heat conduction within a solid wall with fixed surface temperatures at T and $T + \Delta T$, as shown in Figure 3.3. The wall thickness is Δx and its thermal conductivity is k. Determine the thermodynamic force and flow associated with the heat conduction process.

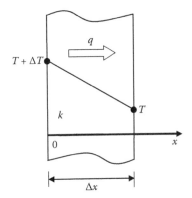

FIGURE 3.3   Irreversibility associated with heat conduction.

SOLUTION

Since the process is steady-state, the entropy generation is equal to the entropy flow across the wall (see Chapter 2):

$$\text{Entropy generation} = - \text{Entropy flow} \tag{3.8}$$

Denoting heat flux by q, the equality in Equation (3.8) can be formulated as

$$\sigma\, x = -\left(\frac{q}{T + T} - \frac{q}{T} \right) \tag{3.9}$$

Rearranging Equation (3.9) results in

$$\sigma = q\; \frac{-\left(\frac{1}{T + T} - \frac{1}{T} \right)}{x} \tag{3.10}$$

The limiting case, for a differential element, dx can be obtained by letting $\Delta x \rightarrow 0$. The result is

$$\sigma = q \frac{d}{dx}\left(\frac{1}{T} \right) \tag{3.11}$$

Therefore, the thermodynamic force and the thermodynamic flux in a heat conduction process are

$$X = \frac{d}{dx}\left(\frac{1}{T} \right) \tag{3.12}$$

$$J = q$$

The result of this steady-state, one-dimensional heat conduction can be generalized to multidimensional heat conduction problems as

$$\sigma = \vec{q} \cdot \quad \left(\frac{1}{T} \right) \tag{3.13}$$

where $\vec{q}$ is the heat flux vector. It is interesting to note that the thermodynamic force associated with heat flow q is not the gradient of temperature, but the gradient of inverse temperature. ▲

TABLE 3.1

Thermodynamic Fluxes and Their Conjugate Forces

Dissipative Process	Generalized Force, X	Generalized Flow, J
Heat conduction	Temperature gradient, $\nabla(1/T)$	Heat flux, $\boldsymbol{q}$
Plastic deformation of solids	Stress, σ/T	Plastic strain, $\dot{\varepsilon}_p$
Damage in solids	Internal variables affinity, A_k/T	Internal variables, $\dot{V}_k$
Chemical reaction [a]	Reaction affinity, A/T	Reaction velocity, v
Mass diffusion [b]	Chemical potential, $-\nabla(\mu_k/T)$	Diffusion flux, $\boldsymbol{J}_k$
Viscous dissipation	Velocity gradient, $\nabla(\boldsymbol{u}/T)$	Shear stress, τ
Joule effect	Voltage, V/T	Electric current, I_e

[a] A denotes the chemical affinity.

[b] μ denotes the chemical potential.

3.1.1 Examples of Thermodynamic Forces and Flows

Thermodynamic forces and flows for other irreversible processes can be defined in a similar way as the foregoing example. For instance, the presence of an electrical potential difference VI_e across the wall of Example 3.1 causes an electric current I_e to flow through the wall. The corresponding entropy generation associated with the flow of electric current is

$$\sigma = I_e \cdot \frac{V}{T} \tag{3.14}$$

with thermodynamic force of $X = V/T$ and thermodynamic flow of $J = I_e$. Voltage acts as the driving force that causes current to flow. Table 3.1 shows explicit expressions for thermodynamic forces and fluxes for some irreversible processes.

3.2 RELATIONS BETWEEN THERMODYNAMIC FORCES AND FLOWS

As mentioned in the foregoing discussion, X_i are the driving forces for thermodynamic fluxes J_i. For instance, as we illustrated in Example 3.1, the presence of temperature gradient drives the flow of heat across the wall. In general, n dissipative processes may be simultaneously involved so that thermodynamic fluxes J_i are functions of their conjugate forces X_i as well as all other driving forces X_j $(j \neq i)$ and temperature T (Bejan 1988), that is,

$$J_i = J_i (X_1,..., X_n, T) \ (i = 1, ..., n) \tag{3.15}$$

Let us assume, for simplicity, that thermodynamic flux is only a function of driving forces (Ziegler 1983). Further, assume that there are only two dissipative processes ($n = 2$) present, so the thermodynamic fluxes are simply:

$$J_1 = J_1 (X_1, X_2) \tag{3.16}$$

$$J_2 = J_2 (X_1, X_2)$$

The question that arises is: How does one determine the function that describes the relation between thermodynamic forces and flows in Equation (3.16)? Considering

Example 3.1, it is obvious that if the temperature gradient across the wall vanishes, the heat flux must vanish as well. This implies that the thermodynamic fluxes J_i vanish if all the thermodynamic forces are zero. This is particularly relevant to equilibrium state in which the driving forces are zero and both forces and flows vanish (Glansdorff and Prigogine 1971):

$$X_i = 0, J_i = 0 \tag{3.17}$$

and entropy generation is nil, $\sigma = 0$, so that the process is nondissipative. The substantiation of this concept is given in Example 3.2. However, Rajagopal and Srinivasa (2004) point out that although the entropy production vanishes at equilibrium state, both conditions in Equation (3.17) may not necessarily apply. More specifically, for some materials such as granular materials (with frictional behavior) and most metals and polymers (with yielding behavior), it is not necessary for all thermodynamic forces X_i to vanish at equilibrium state, and they can, in fact, be nonzero. Conversely, if some of the thermodynamic fluxes J_i are nonzero, the process is strictly dissipative; that is, the entropy generation is nonzero. It is, therefore, the presence of fluxes that determine the dissipative trait of the process.

We now begin our presentation of the relationship between thermodynamic forces and flows by studying a special case of linear processes. Later in this section, we generalize the results by applying the *Thermodynamic Orthogonality Principle* introduced by Ziegler (1983).

Example 3.2

Consider a system involving a set of two thermodynamic forces and flows (X_1, J_1) and (X_2, J_2). Derive the appropriate expressions for fluxes in terms of forces for the system near the equilibrium state.

SOLUTION

We start with Equation (3.16) representing fluxes as a function of forces:

$$J_1 = J_1 (X_1, X_2) \tag{3.18}$$

$$J_2 = J_2 (X_1, X_2)$$

Expanding J_1 in Taylor series near the equilibrium state, we have

$$J_1 = (J_1)_{eq} + \frac{\partial J_1}{\partial X_1} X_1 + \frac{\partial J_1}{\partial X_2} X_2 + \frac{1}{2} \frac{\partial^2 J_1}{\partial (X_1)^2} (X_1)^2 + \frac{1}{2} \frac{\partial^2 J_1}{\partial (X_2)^2} (X_2)^2 + \cdots \tag{3.19}$$

The first term on the right-hand side of Equation (3.19) is zero, that is, $(J_1)_{eq} = 0$. Since the system is near equilibrium state, we merely keep the first-order terms in Equation (3.19) and neglect the higher-order terms. Therefore,

$$J_1 \cong \frac{\partial J_1}{\partial X_1} X_1 + \frac{\partial J_1}{\partial X_2} X_2 = L_{11}X_1 + L_{12}X_2 \tag{3.20}$$

where

$$L_{11} = \frac{\partial J_1}{\partial X_1} \tag{3.21a}$$

$$L_{12} = \frac{\partial J_1}{\partial X_2} \tag{3.21b}$$

The quantities L_{11} and L_{12} are the so-called phenomenological coefficients. They are assumed to be constant and independent of fluxes and forces. The coefficient L_{11} determines the effect of force X_1 on its conjugate flux J_1, and the coefficient L_{12} determines the effect of force X_2 as it drives its cross flux J_1. This is known as *cross coupling effect*. For example, in a plastic deformation of a solid, not only does the force $(1/T)$ cause the flow of heat, $\mathbf{q}$, but also it can drive the plastic strain rate, $\dot{\varepsilon}_p$.

In a similar fashion, the expression for J_2 can be defined as

$$J_2 \cong \frac{\partial J_2}{\partial X_1} X_1 + \frac{\partial J_2}{\partial X_2} X_2 = L_{21}X_1 + L_{22}X_2 \tag{3.22}$$

where

$$L_{21} = \frac{\partial J_2}{\partial X_1} \tag{3.23a}$$

$$L_{22} = \frac{\partial J_2}{\partial X_2} \tag{3.23b}$$

Entropy generation associated with this set of forces and fluxes can be determined from Equation (3.7):

$$\sigma = L_{11}(X_1)^2 + (L_{12} + L_{21})\, X_1X_2 + L_{22}(X_2)^2 \tag{3.24}$$

In general, when there are n pairs of thermodynamic forces and flows in the system, the entropy generation can be written as

$$\sigma = L_{ij} X_i X_j \ (i,\, j = 1,\, 2,\, 3,\ldots,\, n) \tag{3.25}$$

Note that the entropy generation in Equation (3.25) takes the form of a quadratic function for linear systems. ▲

The linear systems are observed for a wide range of experimental conditions (De Groot and Mazur 1962). In this special case, the well-established *Onsager reciprocal relations*[*] between phenomenological coefficients read

$$L_{ij} = L_{ji} \ \forall\, i \neq j \tag{3.26}$$

These relationships are valid for systems near equilibrium where fluxes and forces both deviate by small amounts from their equilibrium state. In fact, the Onsager reciprocal relations provide a powerful tool to study coupling effect in several important phenomena such as thermoelectricity, and thermoelectromagnetic, diffusion, and chemical reactions.

---

[*] The reciprocal relations were first established in 1931 by Lars Onsager who was awarded the Nobel Prize in 1968 for this discovery. These are fundamental relationships for the thermodynamics of irreversible processes.

To elaborate, let us consider the entropy generation due to simultaneous application of electric current I_e and heat flow q. Using the information given in Table 3.1 and employing Equation (3.25), the entropy generation for one-dimensional analysis is written as

$$\sigma = -\frac{q}{T^2}\frac{dT}{dx} + I_e\frac{V}{T} \tag{3.27}$$

Considering Equation (3.20), the thermodynamic fluxes can be expressed as

$$q = -\frac{L_{11}}{T^2}\frac{dT}{dx} + L_{12}\frac{V}{T} \tag{3.28}$$

$$I_e = -\frac{L_{21}}{T^2}\frac{dT}{dx} + L_{22}\frac{V}{T} \tag{3.29}$$

where phenomenological coefficients L_{ij} are determined experimentally. L_{11} and L_{22} can be easily measured from Fourier's law and Ohm's law, respectively:

$$L_{11} = kT^2 \tag{3.30}$$

$$L_{22} = \frac{T}{r} \tag{3.31}$$

where k is the thermal conductivity and r represents the electric resistance. To measure the cross coefficient L_{21}, an experiment can be performed by providing a temperature difference between two junctions of dissimilar metals (Miller 1960) which produces an electric potential between two junctions known as the *Seebeck effect*.[*] This potential is measured at zero current. Therefore, by setting $I_e = 0$, from Equation (3.29) we can write

$$L_{21} = \frac{VT^2}{rdT/dx} \tag{3.32}$$

In another experiment, a current I_e is provided while two junctions are kept at constant temperature. This results in a flow of heat between two junctions widely known as the *Peltier effect*.[†] Since temperature is uniform $dT/dx = 0$, Equations (3.28) and (3.29) result in

$$L_{12} = \frac{qT}{rI_e} \tag{3.33}$$

Experimental measurements (Miller 1960) of the right-hand sides of Equations (3.32) and (3.33) show that

$$\frac{VT^2}{rdT/dx} = \frac{qT}{rI_e} \tag{3.34}$$

which verifies the Onsager reciprocal relations:

$$L_{12} = L_{21} \tag{3.35}$$

---

[*] The Seebeck effect was discovered by Thomas J. Seebeck in 1821.
[†] The Peltier effect was discovered by Jean C. Peltier in 1834.

3.2.1 Thermodynamic Orthogonality Principle

It is important to point out that for some irreversible processes, such as plastic deformation in solids, the assumption asserting that forces and flows are linearly related is invalid and consideration of Onsager's reciprocal relation is, in fact, inapt. Indeed, complications may arise since the phenomenological coefficients, L_{ij}, may depend on fluxes. To treat these cases, we employ the principle of thermodynamic orthogonality introduced by Ziegler (1983). Hans Ziegler (1983) established the so-called Thermodynamic Orthogonality Principle (TOP) for the thermodynamic forces in dissipative processes. He demonstrated the usefulness of TOP in setting up constitutive equations between forces and fluxes. Ziegler's principle is truly a powerful technique for analyzing processes involving nonequilibrium thermodynamics. In fact, Ziegler has shown that the second law of thermodynamics, which is supplemented with statistical arguments due to Boltzmann and Gibbs, can be obtained as a corollary of TOP. Interestingly, Martyushev and Seleznev (2006) proved that Onsager's reciprocity principle can be deduced from TOP subjected to assumption of linearity.

The orthogonality principle simply states that for a given thermodynamic flux, J_i, its conjugate thermodynamic force X_i is normal to the dissipation surface Φ = const. Here, for brevity in notation, we consider the entropy generation function σ instead of dissipation function Φ as they are related by $\Phi = T\sigma$. Let us consider a system involving only one pair of thermodynamic force and flow (X, J). The thermodynamic flux is defined by vector J as shown in Figure 3.4. The entropy generation function associated with the flux vector is given by σ = const. Therefore, the magnitude of the thermodynamic force can be evaluated from Equation (3.7) as $|X| = \sigma/|J|$. However, its direction is determined via the orthogonality condition as shown in Figure 3.4. The thermodynamic force X corresponding to the flow J is perpendicular to the entropy generation function in the end point of J (point O in Figure 3.4). Therefore, TOP asserts the following relation for determining the driving force:

$$X = M \frac{\partial \sigma}{\partial J} \tag{3.36a}$$

where the proportionality coefficient M is obtained using Equation (3.7):

$$M = \frac{\sigma}{(\partial \sigma / \partial J)J} \tag{3.36b}$$

Equations (3.36a) and (3.36b) are known as the orthogonality conditions signifying that the force X corresponding to flux J is orthogonal to the surface $\sigma(J)$ = const.

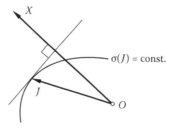

FIGURE 3.4   The orthogonality condition for the force X.

The orthogonality condition in Equations (3.36a) and (3.36b) is given for only one flux. When there are n pairs of thermodynamic forces and flows in the system, the orthogonality condition can be written as

$$X_i = M \frac{\partial \sigma}{\partial J_i} \quad (i = 1, 2, 3, \ldots, n) \tag{3.37a}$$

$$M = \frac{\sigma}{(\partial \sigma / \partial J_i) J_i} \tag{3.37b}$$

The orthogonality principle discussed above is formulated in the flux space as the entropy generation depends only on X_i. Alternatively, the principle can be formulated in the force space when entropy generation depends only on J_i. In that case, one can simply rewrite Equations (3.36a) and (3.36b) substituting J for X.

To gain insight into the understanding of the orthogonality conditions, we discuss its derivation through the following example for a simple case of one flux, J.

Example 3.3

Consider a system consisting of one pair of the thermodynamic force and flux (X, J). The entropy generation is given as $\sigma = \sigma(J)$. Starting from the power series expansion of entropy generation in the vicinity of J, derive the orthogonality principle as given in Equation (3.36a) and Equation (3.36b).

SOLUTION

Expanding $\sigma(J)$ in the Taylor series in the vicinity of J yields

$$\sigma(J + \delta J) = \sigma(J) + \frac{\partial \sigma}{\partial J} \delta J + \frac{1}{2} \frac{\partial^2 \sigma}{\partial J^2} (\delta J)^2 + \cdots \tag{3.38}$$

The term $\partial \sigma / \partial J$ is the only term in the expansion of Equation (3.38) that has directionality (i.e., it is a vector). Hence, it can determine the direction of dissipative force X. We now need to correlate $\partial \sigma / \partial J$ with X. Starting from Equation (3.7):

$$\sigma(J + \delta J) = X(J + \delta J)(J + \delta J) \tag{3.39}$$

Similarly, X can be expanded in power series:

$$X(J + \delta J) = X(J) + \frac{\partial X}{\partial J} \delta J + \frac{1}{2} \frac{\partial^2 X}{\partial J^2} (\delta J)^2 + \cdots \tag{3.40}$$

Substitution of Equation (3.40) into Equation (3.39) and neglecting the high-order terms yields

$$\sigma(J + \delta J) = XJ + \left(X + J \frac{\partial X}{\partial J} \right) \delta J \tag{3.41}$$

Substituting $\sigma(J + \delta J)$ from Equation (3.41) into Equation (3.38) gives

$$\sigma(J) + \frac{\partial \sigma}{\partial J}\delta J = XJ + \left(X + J\frac{\partial X}{\partial J} \right)\delta J \tag{3.42}$$

Recalling that $\sigma(J) = XJ$, the first terms on both sides of the Equation (3.42) cancel out and we have

$$\frac{\partial \sigma}{\partial J} = X\left(1 + \frac{J}{X}\frac{\partial X}{\partial J} \right) \tag{3.43}$$

The term in the parentheses can be written as follows:

$$1 + \frac{J}{X}\frac{\partial X}{\partial J} = 1 + \frac{J^2}{\sigma}\left(\frac{1}{J}\frac{\partial \sigma}{\partial J} - \frac{\sigma}{J^2} \right) = \frac{J}{\sigma}\frac{\partial \sigma}{\partial J} \tag{3.44}$$

Substitution of Equation (3.44) into Equation (3.43) and rearranging the resultant:

$$X = M\frac{\partial \sigma}{\partial J} \tag{3.45a}$$

where M is given by

$$M = \frac{\sigma}{(\partial \sigma / \partial J)J} \tag{3.45b}$$

Equations (3.45a) and (3.45b) describe the orthogonality principle given by Equations (3.36a) and (3.36b). Alternative proof of the orthogonality principle is available in the literature employing the variational principle. Interested readers can refer, for example, to Ziegler (1983) and Martyushev and Seleznev (2006). ▲

It is to be noted that the preceding derivations are given for a pair of vectors $\{X, J\}$, such as $\{\nabla(1/T), q\}$. Ziegler (1983) showed that the orthogonality conditions still hold if fluxes and forces are defined by second-order tensors such as $\{\sigma, \dot{\varepsilon}_p\}$.

The fluxes characterizing a fatigue process involve a set of two tensors (e.g., the heat flow vector and the rate of plastic strain tensor). The entropy generation depends on all fluxes, and hence, forces depend on their conjugate fluxes as well as all other fluxes. Such a dissipative process is referred to as *coupled*. However, if forces only depend on their counterpart fluxes, the process is said to be *uncoupled*.

The orthogonality conditions provide a powerful tool for assessment of thermodynamic forces. For example, recall that the most common form of the entropy generation σ is a positive definite quadratic function (Equation (3.25)):

$$\sigma = L_{ij}X_i X_j \quad (i, j = 1, 2, 3, \ldots, n) \tag{3.46}$$

which reduces to the following when two fluxes are involved (Equation (3.24)):

$$\sigma = L_{11}(X_1)^2 + (L_{12} + L_{21})X_1 X_2 + L_{22}(X_2)^2 \tag{3.47}$$

Applying the orthogonality conditions of Equation (3.36a) and Equation (3.36b) in the force space results in the following:

$$M = \frac{\sigma}{(\partial\sigma/\partial X_i)X_i} = L_{ki}\frac{X_iX_k}{2L_{ki}\ X_iX_k} = \frac{1}{2} \tag{3.48a}$$

$$J_1 = \frac{1}{2}(2L_{11}X_1 + 2L_{12}X_2) = L_{11}X_1 + L_{12}X_2 \tag{3.48b}$$

$$J_2 = \frac{1}{2}(2L_{12}X_1 + 2L_{22}X_2) = L_{12}X_1 + L_{22}X_2 \tag{3.48c}$$

Therefore, the quadratic dissipation function results in linear constitutive equations for forces as shown by Equation (3.20). The coefficient L_{12} describes the cross-effect between the two dissipative processes X_1 and X_2.

3.2.2 COUPLING BETWEEN PLASTIC DEFORMATION AND HEAT FLOW

Many attempts have been made during the past decades to underpin the scientific investigation for the coupling between plastic deformation and heat flow in materials. Some of the major studies within the context of thermodynamics of irreversible processes in materials have been carried out by Kluitenberg (1962a–c, 1963, 1977) and Kluitenberg, De Groot, and Mazur (1953a,b, 1955). Kluitenberg (1962c) derived the equation for entropy generation (dissipation function) for a media subjected to plastic flow, viscous flow, and heat conduction. He developed the phenomenological equations for thermodynamic forces and fluxes when media is either isotropic or anisotropic and concluded that for the case of anisotropic medium, there may exist a coupling effect among all these dissipative phenomena. However, in isotropic materials such a coupling effect does not exist between heat conduction and plastic flow.

Hackl, Fischer, and Svoboda (2010) presented an interesting numerical example to illustrate the cross-effect between plastic deformation and heat conduction for steel. They postulated that the decoupling of the elementary dissipative processes such as heat flow and deformation is possible only if the dissipation potential (entropy generation) is a homogenous function of the same order for all fluxes (e.g., the heat flux and the rate of plastic strain); otherwise, the process is coupled. Taking into account the orthogonality condition, they put forward formulations for underpinning the coupling of processes in which the dissipation function takes a mathematical form. Particular attention is given to the coupling of heat flux q and the rate of plastic deformation $\dot{\varepsilon}_p$. Without going into details, here we quote the resulting equation of their one-dimensional analysis for the coupling of heat flux and rate-independent plasticity as

$$q^2\left(\frac{1}{2}TR^2q + R\ \ T\right) + \sigma_0\dot{\varepsilon}_p(\ \ T + TRq) = 0 \tag{3.49}$$

where σ_0 is the yield stress and $R = T/(MTq)$ with M obtained from Equation (3.47). For the limiting case of $\dot{\varepsilon}_p = 0$, the terms in the first parenthesis vanish. Hence:

$$q = -2\frac{T}{TR} \quad \text{for} \quad \dot{\varepsilon}_p \to 0 \tag{3.50a}$$

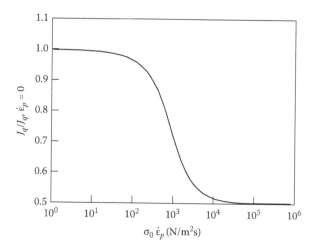

FIGURE 3.5 Variation of heat flux due to coupling effect. Material properties and conditions are $\sigma_0 = 3 \times 10^7$ N/m², $T = 300$ K, $\nabla T = 100$ K/m, and $R = 1.5 \times 10^{-4}$ s/N. (Reproduced from Hackl, K., Fischer, F.D., and Svoboda, J., *Proc. R. Soc. A* 467, 1186–1196, 2010.)

And for the limiting case of $\dot{\varepsilon}_p \rightarrow \infty$ the terms in the second parentheses vanish, and the following equation results

$$q = -\frac{T}{TR} \quad \text{for} \quad \dot{\varepsilon}_p \rightarrow \infty \tag{3.50b}$$

Comparison of Equations (3.50a) and (3.50b) reveals that the heat flux in the presence of plastic deformation reduces to a half due to the coupling effect. Hackl, Fischer, and Svoboda (2010) conclude that "the microstructure, which has changed owing to plastic deformation, retards the heat flux, which is in accordance with physics of phonons as carriers of heat." It is shown that defects, dislocations, and impurities in the material cause scattering of phonons (Goodson and Flik 1993; Gruner and Bross 1968). Figure 3.5 shows the variation of heat flux as a function of plastic strain rate. The material properties are given in the caption of the figure.

So far, we have studied the fundamental concepts of thermodynamic forces and flows necessary to characterize a fatigue process. In the following section, we introduce the Degradation–Entropy Generation (DEG) theorem originally developed by Bryant, Khonsari, and Ling (2008) and present its application to processes involving cyclic fatigue.

3.3 THE DEGRADATION–ENTROPY GENERATION THEOREM

The benefit of employing thermodynamic forces and flows is that the entropy production σ can be explicitly expressed in terms of experimentally measurable quantities. As inferred by the second law of thermodynamics, during an irreversible process, the available work dissipates to a lower-quality energy in the form of heat. It follows, therefore, that in the course of a dissipative process, the high-quality energy degrades to the low-quality energy and that the measure of the degradation is entropy generation. However, in this book, we refer to degradation in a mechanical sense as a measure of *damage* and exhaustion of the system's lifespan. In literature, the terminology of *aging* has also been used, referring to the loss of integrity of the system during the course of dissipative processes. In the remaining part of this chapter, we introduce the concepts of degradation forces and flows analogous to thermodynamic forces and flows followed by a discussion on a general theorem known as the *Degradation–Entropy Generation* (DEG) theorem.

3.3.1 DEGRADATION FORCES AND FLOWS

Let us start with considering a system in which only one dissipative process p is responsible for causing degradation. Assume that the dissipative process $p = p(\zeta)$ depends on a time-dependent phenomenological variable ζ, for example, plastic strain in a process involving fatigue. The entropy production of the system $d_i s$ depends on dissipative processes p, and its rate $\sigma = d_i s/dt$ can be obtained using the chain rule:

$$\sigma = \frac{d_i s}{dt} = \left(\frac{\partial_i s}{\partial p} \frac{\partial p}{\partial \zeta} \right) \frac{\partial \zeta}{\partial t} = XJ \tag{3.51}$$

where

$$X = \frac{\partial_i s}{\partial p} \frac{\partial p}{\partial \zeta}$$

$$J = \frac{\partial \zeta}{\partial t}$$

which is the expression obtained for entropy generation in Equation (3.7) with X and J representing the thermodynamic forces and flows, respectively. Note that here p is the only dissipative process responsible for entropy generation, and it is the only source that causes degradation in the system. Now, let w represent a measure of the system's degradation. Clearly, w depends on dissipative process, that is, $w = w(p)$. Analogous to the entropy generation, the rate of degradation $D = dw/dt$ can be obtained using the chain rule as

$$D = \frac{dw}{dt} = \left(\frac{\partial w}{\partial p} \frac{\partial p}{\partial \zeta} \right) \frac{\partial \zeta}{\partial t} = YJ \tag{3.52}$$

where

$$Y = \frac{\partial w}{\partial p} \frac{\partial p}{\partial \zeta}$$

$$J = \frac{\partial \zeta}{\partial t}$$

In conjunction with thermodynamic force, Bryant, Khonsari, and Ling (2008) refer to Y as the degradation force. It is to be noted that the degradation of the system depends on

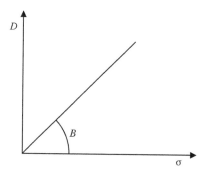

FIGURE 3.6 Linear relationship between rate of degradation parameter and entropy generation.

the same dissipative process p as does the entropy generation. Considering the fact that thermodynamic flow J is a joint parameter in Equations (3.51) and (3.52), a degradation coefficient can be defined as

$$B = \frac{Y}{X} = \frac{(\partial w/\partial p)(\partial p/\partial \zeta)}{(\partial_i s/\partial p)(\partial p/\partial \zeta)} = \frac{\partial w}{\partial_i s}\Big|_p \tag{3.53}$$

Equation (3.53) suggests that B measures how entropy generation and degradation interact on the level of dissipative process p. Figure 3.6 schematically shows the relation between rate of degradation D and rate of entropy generation σ as correlated with degradation coefficient B.

The DEG theorem can be easily generalized for a system consisting of N dissipative processes, p_n ($n = 1, 2, 3, \ldots N$). In this case, the overall entropy generation is the sum of all entropy generations by each process (Bryant, Khonsari, and Ling 2008):

$$\sigma = \sum_n \sum_m \left(\frac{\partial_i s}{\partial p_n} \frac{\partial p_n}{\partial \zeta_n^m} \right) \frac{\partial \zeta_n^m}{\partial t} = \sum_n \sum_m X_n^m J_n^m \tag{3.54}$$

Note that each dissipative process $p_n = p_n(\zeta_n^m)$ depends on a set of time-dependent phenomenological variables ζ_n^m ($m = 1, 2, 3, \ldots M$). Similarly, the degradation rate in Equation (3.52) can be presented in a general form as

$$D = \sum_n \sum_m \left(\frac{\partial w}{\partial p_n} \frac{\partial p_n}{\partial \zeta_n^m} \right) \frac{\partial \zeta_n^m}{\partial t} = \sum_n \sum_m Y_n^m J_n^m \tag{3.55}$$

where X_n^m are the thermodynamic forces and Y_n^m are the degradation forces in a system consisting of more than one dissipative process.

3.3.2 GENERALIZATION: DEG COROLLARY

The degradation parameter w or its rate $D = dw/dt$ can account for different modes of degradation depending on the problem under study. For example, in sliding contact problems where friction is a degrading mechanism, the degradation parameter w is defined as volume of material removal, i.e., wear. In a process involving fatigue, the degradation parameter can be defined as crack length or damage parameter as popularly used in *Continuum Damage Mechanics* (CDM) which is covered in Chapter 7. Here, let us elaborate more on an important corollary of the DEG theorem.

From Equation (3.51) through Equation (3.53), it is inferred that the degradation rate D is a *linear combination* $D = B\sigma$ of the components of entropy production σ of the dissipative process p.

The generalized degradation force Y is a linear function $Y = BX$ of the generalized thermodynamic force X, and the proportionality factor B is the degradation coefficients given by $B = \partial w/\partial_i s|_p$. The integration of Equation (3.52) over time suggests that the total degradation

$$\int D = B \int \sigma \qquad (3.56)$$

is a linear combination of the total *accumulated entropy* $\int \sigma$ generated by the dissipative process p. If a critical value of degradation $(\int D)_{cr}$ exists, which could result in system failure, to satisfy this last equation, corresponding critical values of accumulated irreversible entropy generation $(\int \sigma)_{cr}$ must also exist. This corollary of DEG theorem suggests a very interesting and promising approach to assess time to failure in fatigue problems. Chapter 5 and 6 are devoted to this subject.

3.3.3 Application: Paris–Erdogan Law

Let us turn now our attention to application of these concepts to a typical cyclic fatigue loading. In the following, we discuss how the DEG theorem can be used to assess crack propagation rate and how the well-known Paris–Erdogan law of crack propagation can be deduced by using the DEG theorem (Amiri and Khonsari 2012).

Consider a crack of length a growing at a steady rate (see Figure 3.7). We define the degradation parameter to be the crack length, that is, $w = a$. Therefore, degradation can be defined as $a = a\{W_p(N)\}$, where the plastic energy generation at the crack tip, W_p, is the dominant dissipative process with number of cycles N as the phenomenological variable. Using the DEG theorem, we can derive a formula describing the rate of crack propagation.

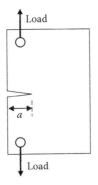

FIGURE 3.7 Schematic of crack propagation.

Starting from Equations (3.51) and (3.52), we can write

$$d_i s/dt = XJ \text{ and } da/dt = YJ \tag{3.57}$$

where $J = dN/dt = f$ (f is the test frequency), $X = (d_i s/dW_p)(dW_p/dN)$ and $Y = (da/dW_p) \times (dW_p/dN)$.

Let us assume that the plastic energy is dissipated as entropy generation, that is, $dW_p = Td_i s$. Therefore, the entropy production can be written as

$$\sigma = \frac{d_i s}{dt} = \frac{\partial_i s}{\partial p} \frac{\partial p}{\partial N} \frac{\partial N}{\partial t} = \frac{f}{T} \frac{dW_p}{dN} \tag{3.58}$$

where $X = (1/T)(dW_p/dN)$. Applying Equation (3.52) yields

$$D = \frac{da}{dt} = YJ = BXJ = B\frac{f}{T}\frac{dW_p}{dN} \tag{3.59}$$

The degradation coefficient B in Equation (3.59) measures how entropy generation and crack propagation interact on the level of dissipative plastic deformation process.

To deduce the Paris law of crack propagation, we use methods developed to assess the energy dissipation in the plastic zone ahead of the crack tip W_p. Pertinent works can be found in number of research papers (see, for example, Mura and Lin 1974; Irving and Mccartney 1977; Liaw, Kwun, and Fine 1981; Bodner, Davidson, and Lankford 1983; Klingbeil 2003). Here, we borrow a correlation presented by Bodner, Davidson, and Lankford (1983) in the following form:

$$\frac{dW_p}{dN} = At\frac{(\Delta K)^4}{\mu \sigma_y^2} \tag{3.60}$$

where A is a dimensionless constant, t is the specimen thickness, ΔK is the stress intensity factor, μ is the shear modulus, and σ_y is the yield stress. Readers interested in further detail on dimensionless constant A can refer to the comprehensive work of Liaw, Kwun, and Fine (1981).

Substitution of Equation (3.60) into Equation (3.59) yields

$$\frac{da}{dN} = \frac{da}{fdt} = B\frac{At}{T}\frac{(\Delta K)^4}{\mu \sigma_y^2} = C(\Delta K)^4 \tag{3.61}$$

Equation (3.61) represents the Paris–Erdogan law (1963) of crack propagation with constant C defined as

$$C = B\frac{At}{T\mu \sigma_y^2} \tag{3.62}$$

Equation (3.62) signifies the intimate relation between the constant C in Paris–Erdogan law and the degradation coefficient B, which is, in turn, related to entropy generation via

Equation (3.58). This implies that the rate of crack propagation is determined by the intensity of the degradation coefficient and that the empirical Paris–Erdogan law can be viewed as a consequence of the DEG theorem (Amiri and Khonsari 2012). This result further illustrates the usefulness of the entropy approach to the fatigue problems.

It is worth mentioning that the DEG theorem provides a technique to assess degradation in a variety of applications. Bryant, Khonsari, and Ling (2008), for example, demonstrated that the well-known Archard's law for wear prediction can be deduced from the DEG theorem. However, its application in other disciplines provides an interesting research direction for years to come.

REFERENCES

Amiri, M. and Khonsari, M.M. 2012. On the role of entropy generation in processes involving fatigue. *Entropy* 14, 24–31.

Annamalai, K. and Puri, I.K. 2002. *Advanced Thermodynamics Engineering*. Boca Raton, FL: CRC Press.

Bejan, A. 1988. *Advanced Engineering Thermodynamics*. New York: John Wiley & Sons, Inc.

Bodner, S.R., Davidson, D.L., and Lankford, J. 1983. A description of fatigue crack growth in terms of plastic work. *Eng. Frac. Mech.* 17, 189–191.

Dai, Z. and Xue, Q. 2009. Progress and development in thermodynamic theory of friction and wear. *Sci. China Ser E-Tech. Sci.* 52, 844–849.

De Groot, S.R. and Mazur, P. 1962. *Non-Equilibrium Thermodynamics*. New York: Interscience Publishers.

Glansdorff, P. and Prigogine, I. 1971. *Thermodynamic Theory of Structure, Stability and Fluctuations*. New York: John Wiley & Sons, Inc.

Goodson, K.E. and Flik. M.I. 1993. Electron and phonon thermal conduction in epitaxial high-Tc superconducting films. *ASME J. Heat Transfer* 115, 1–25.

Gruner, P. and Bross, H. 1968. Limitation of the phonon heat conductivity by edge-dislocation dipoles, single dislocations, and normal processes. *Phys. Rev.* 172, 583–597.

Hackl, K., Fischer, F.D., and Svoboda, J. 2010. A study on the principle of maximum dissipation for coupled and non-coupled non-isothermal processes in materials. *Proc. R. Soc. A* 467, 1186–1196.

Irving, P.E. and Mccartney, L.N. 1977. Prediction of fatigue crack growth rates: Theory, mechanisms, and experimental results. *Metal. Sci.* 11, 351–361.

Jou, D., Casas-Vazquez, J., and Lebon, G. 2010. *Extended Irreversible Thermodynamics*, 4th ed. Berlin and Heidelberg: Springer-Verlag GmbH & Co.

Klamecki, B.E. 1980a. Wear—an entropy production model. *Wear* 58, 325–330.

Klamecki, B.E. 1980b. A thermodynamic model of friction. *Wear* 63, 113–120.

Klamecki, B.E. 1982. Energy dissipation in sliding. *Wear* 77, 115–128.

Klamecki, B.E. 1984. Wear—an entropy based model of plastic deformation energy dissipation in sliding. *Wear* 96, 319–329.

Klingbeil, N.W. 2003. A total dissipated energy theory of fatigue crack growth in ductile solids. *Int. J. Fatigue* 25, 117–128.

Kluitenberg, G.A. 1962a. A note on the thermodynamics of Maxwell bodies, Kelvin bodies (Voigt bodies), and fluids. *Physica* 28, 561–568.

Kluitenberg, G.A. 1962b. On rheology and thermodynamics of irreversible processes. *Physica* 28, 1173–1183.

Kluitenberg, G.A. 1962c. Thermodynamical theory of elasticity and plasticity. *Physica* 28, 217–232.

Kluitenberg, G.A. 1963. On the thermodynamics of viscosity and plasticity. *Physica* 29, 633–652.

Kluitenberg, G.A. 1977. A thermodynamic discussion of the possibility of singular yield conditions in plasticity theory. *Physica A* 88, 122–134.

Kluitenberg, G.A. and De Groot, S.R. 1954a. Relativistic thermodynamics of irreversible processes. IV: Systems with polarization and magnetization in an electromagnetic field. *Physica* 21, 148–168.

Kluitenberg, G.A. and De Groot, S.R. 1954b. Relativistic thermodynamics of irreversible processes. V: The energy momentum tensor of the macroscopic electromagnetic field, the macroscopic forces acting on the matter and the first and second laws of thermodynamics. *Physica* 21, 169–192.

Kluitenberg, G.A., De Groot, S.R., and Mazur, P. 1953a. Relativistic thermodynamics of irreversible processes. I: Heat conduction, diffusion, viscous flow and chemical reactions; Formal part. *Physica* 19, 689–704.

Kluitenberg. 1953b. Relativistic thermodynamics of irreversible processes, II: Heat conduction and diffusion; Physical part. *Physica* 19, 1079–1794.

Lebon, G., Jou, D., and Casas-Vazquez, J. 2008. *Understanding Non-equilibrium Thermodynamics: Foundations, Applications, Frontiers*. Berlin and Heidelberg: Springer-Verlag.

Liaw, P.K., Kwun, S.I., and Fine, M.E. 1981. Plastic work of fatigue crack propagation in steels and aluminum alloys. *Metall. Trans. A* 12, 49–55.

Martyushev, L.M. and Seleznev, V.D. 2006. Maximum entropy production principle in physics, chemistry, and biology. *Phys. Rep.* 426, 1–45.

Miller, D.G. 1960. Thermodynamics of irreversible processes. *Chem. Rev.*, 60, 15–37.

Mura, T. and Lin, C.T. 1974. Theory of fatigue crack growth for work hardening materials. *Int. J. Fracture* 10, 284–287.

Paris, P. and Erdogan, F. 1963. A critical analysis of crack propagation laws. *Trans. ASME Ser D, J. Basic Eng.* 85, 528–534.

Rajagopal K.R. and Srinivasa, A.R. 2004. On thermomechanical restrictions of continua. *Proc. R. Soc. A* 460, 631–651.

Zmitrowicz, A. 1987a. A thermodynamical model of contact, friction and wear: I Governing equations. *Wear* 114, 135–168.

Zmitrowicz, A.1987b. A thermodynamical model of contact, friction and wear: II Constitutive equations for materials and linearized theories. *Wear* 114, 169–197.

Zmitrowicz, A. 1987c. A thermodynamical model of contact, friction and wear: III Constitutive equations for friction, wear and frictional heat. *Wear* 114, 199–221.

Ziegler, H. 1983. *An Introduction to Thermomechanics*. Amsterdam: North-Holland.

4 Fatigue Mechanisms: An Overview

In this chapter, we discuss the fundamentals of fatigue. Particularly, multiaxial and variable amplitude loading fatigue are presented and discussed. Also presented is the relationship between fatigue, energy dissipation, and temperature evolution.

4.1 MULTISCALE CHARACTERISTICS OF FATIGUE

A review of the massive number of publications in the fatigue area reveals that challenges in modeling its mechanisms have been only partially successful: no models can claim to be complete. The difficulties in addressing the degradation in a fatigue system arise from the inherent complexity of the material behavior in response to the load at multiple length scales (Zhang and Wang 2008). Incredibly small devices are now being produced with the representative length scale on the order of micro- and nanometer in which the occurrence of any nanoscale defect during the production could result in sudden failure. Therefore, the applicability of the conventional theoretical and experimental approaches developed for bulk material should be reexamined when dealing with micro- and nanoscale fatigue applications. This calls for the need to address the hierarchical multiscale modeling (Figure 4.1) to complement the traditional approaches. Defects develop initially at the atomic scale due to cyclic slip and grow to a form of microscopic fatigue damage commonly known as microplasticity that accumulates irreversibility to form micro- to macrocracks (Mughrabi 2009). Further application of the cyclic load leads to accumulation and growth of a microcrack(s) into a macrocrack and finally, at the level of macroscale, the crack propagates due to the irreversible plastic deformation at the crack tip until failure occurs.

The next hierarchical level involves extensive component testing and analysis before application in the field can be realized. These involve load-stress analysis and crack-growth and damage analyses with particular attention to the appropriate operating conditions. Growing rapidly in recent years are multiscale computational techniques for characterization of material behavior under deformation; for comprehensive reviews of the state-of-the-art see, for example, Ghoniem et al. (2003), Nosonovsky and Bhushan (2007), and McDowell (2008). In what follows, we review the parameters that influence a fatigue process and discuss different fatigue regimes.

4.2 PARAMETERS INFLUENCING FATIGUE AND CLASSIFICATION OF REGIMES

Figure 4.2 shows a schematic of fatigue failure processes categorized based on the significance of parameters affecting fatigue life. In this figure, parameters that influence fatigue life are categorized into four groups. They include low- and high-cycle fatigue, the state of stress, loading amplitude and loading sequence, as well as testing and environmental conditions. The

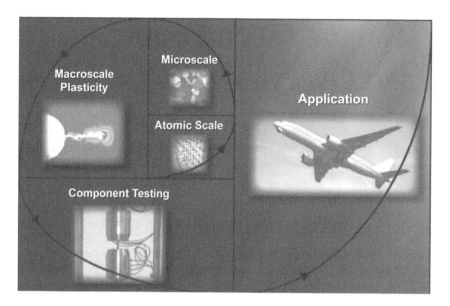

FIGURE 4.1   Hierarchy levels of fatigue damage covering multiscales.

difficulties in addressing the degradation in a fatigue system arise from the inherent complexity of the material behavior in response to different parameters that influence the fatigue mechanisms. For example, the low-cycle fatigue (LCF) mechanism is different from that of the high-cycle fatigue (HCF). In LCF, cracks initiate at the surface of the specimen, where the stresses are maximum, whereas in HCF, cracks nucleate from faults and defects inside the specimen. In what follows, we briefly discuss some of the major class of fatigue problems and their associate challenges.

4.2.1   Low-Cycle Fatigue (LCF) and High-Cycle Fatigue (HCF)

The most common methodology for characterizing the fatigue life is the so-called *S-N* curve or Wöhler diagram which essentially represents the *resistance* of a given material to cyclic application of loads. It describes the observation that the material's strength continuously declines with repeated application of load and that it varies inversely with the number of cycles applied. A typical schematic of an *S-N* curve presented in log-log scale is shown in Figure 4.3 with the applied stress fluctuating between maximum stress S_{max} and minimum stress S_{min}. The ordinate of *S-N* curve represents S_{max}, and the abscissa represents the required number of cycles to failure, N_f. Failure commonly occurs due to pre-existence of micro defects in the material or presence of areas with stress concentration. As a result of applied alternating stress, cracks grow to a critical length so that the cross-section area can no longer carry the applied load and the specimen fractures.

Fatigue tests basically embrace two distinctly different domains of cyclic stress or strain. The first domain is restricted to relatively short fatigue lives, typically less than 10^3 or 10^4 cycles. This domain is referred to as the low-cycle fatigue (LCF). It typically involves large applied stress amplitude, and plastic strain is the dominant dissipation mode. The second domain is associated with fatigue lives greater than 10^6 cycles, wherein stresses and strains are largely confined to the elastic region. This domain is referred to as the high-cycle fatigue

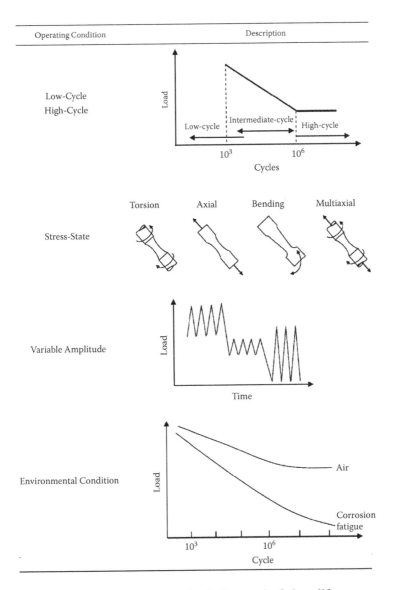

FIGURE 4.2 An overview of the parameters that influence the fatigue life.

(HCF). It is generally associated with low-stress amplitude and long life. The region between 10^3 and 10^6 cycles represents the intermediate-cycle fatigue (ICF). There are also two other regions that can be recognized on the S-N plot: very low-cycle fatigue (VLCF) with number of cycles less than 10^3 and very-high-cycle fatigue (VHCF) with number of cycles greater than 10^8 or 10^9. It is observed that some materials, for example, body centered cubic (BCC) steels (Bannantine, Comer, and Handrock 1997) and titanium alloys (Case, Chilver, and Ross 1999) exhibit an *endurance limit* or *fatigue limit*, S_e, below which the material appears to have an *infinite life*. This is shown by the horizontal line in Figure 4.3. In contrast, nonferrous metals do not have an endurance limit so their S-N curves do not become horizontal (cf. Figure 4.3). Note that the endurance limit can be significantly influenced by several factors particularly when the specimen is exposed to corrosive and/or high temperature environment or when it is experiencing periodic overloads during the course of fatigue. According to

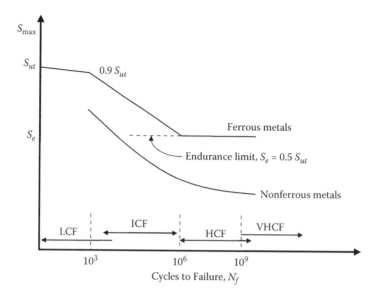

FIGURE 4.3 Fatigue strength of metals as represented by the *S-N* curve.

Bannantine, Comer, and Handrock (1997), for most steels with *ultimate strength*, $S_{ut} < 1380$ MPa, the following relationship between S_{ut} and S_e holds

$$S_e = 0.5\ S_{ut} \quad \text{for} \quad S_{ut} < 1380\ \text{MPa} \tag{4.1}$$

It is also widely accepted that fatigue strength for a life of 10^3 cycles can be estimated as 90% of the ultimate strength as shown in Figure 4.3.

Conventionally, fatigue lives greater than 10^6 cycles are considered to be infinite. Although it is assumed that for stresses below endurance limit, material never fails, experimental observations show that even for very high-cycle fatigue, with stress far below the endurance limit, fatigue failure can, in fact, occur (Bathias and Paris 2004; Lukas and Kunz 1999; Nishijima and Kanazawa 1999; Mughrabi 1999; Szczepanski et al. 2008). For example, Mughrabi (1999) describes the failure mechanism in very high-cycle fatigue processes by bringing into play the role of cyclic slip irreversibility associated with microplastic strains and showing that they accumulate over time and eventually lead to failure. The reader interested in detailed discussion on failure mechanism in high- and very-high-cycle fatigue is referred to Bathias and Paris (2004).

Fatigue tests are generally expensive and time consuming as they may take several days of continuous testing especially for very-high-cycle fatigue. For example, for a single specimen at 10 Hz loading frequency, nearly 27 hrs would be required to amass 10^6 cycles. For a higher number of cycles, experiments become quite costly to operate. Hence, development of models capable of predicting fatigue failure is highly desirable. The well-known Basquin's model assumes a linear correlation between the stress amplitude S and the number of cycles to failure N_f in a log–log scale as (Basquin 1910):

$$S = A(N_f)^B \tag{4.2}$$

where A and B are empirical constants. Equation (4.2) satisfactorily covers a wide range of experiments embracing intermediate-cycle fatigue. Figure 4.4 shows the *S-N* curve for

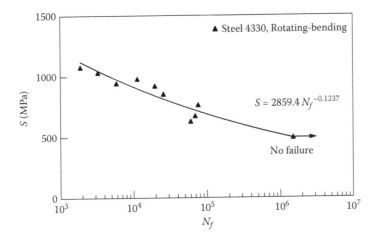

FIGURE 4.4   *S-N* curve of Steel 4330 undergoing rotating-bending fatigue test.

rotating-bending fatigue test of unnotched Steel 4330 specimens at room temperature. Results pertain to completely reversed loading. Figure 4.4 shows that an equation of the form of Equation (4.2) can be fitted to the experimental data with $A = 2859.4$ and $B = -0.1237$. Note that the arrow pointing to the right indicates that failure did not occur under that particular operating condition on these experiments.

As mentioned earlier, fatigue failure is intimately connected to crack initiation and propagation. It is conventionally assumed that in LCF, where stresses are high, failure is rooted in surface-initiated cracks, while in HCF, cracks initiate from inside the material and grow to the surface. However, recent studies show that whether a crack initiates from the surface or subsurface depends on the surface condition of the material being tested. For example, the experimental study of Szczepanski et al. (2008) for very high-cycle fatigue of Ti-6246 in the range of 10^6 to 10^9 cycles reveals that failure occurs either from surface cracks or subsurface cracks. Their results show that as the stress level decreases, there is an increased likelihood of internal crack initiation. Thus, for materials with hardened surface in the HCF regime, failure takes place internally while in the LCF regime, it occurs at the surface (Nishijima and Kanazawa 1999). In conclusion, it can be hypothesized that surface finish and presence of notch and stress concentration are dominant factors in the initiation and propagation of fatigue cracks.

Example 4.1

An axel shaft is made of Steel 4330 with ultimate strength of $S_{ut} = 1240$ MPa. The shaft is subjected to rotating-bending fatigue loading. What would be the alternating applied stress for a finite life of 25×10^3 cycles? Compare the analysis with experimental data given in Figure 4.4.

SOLUTION

Since the ultimate strength of the material is below 1380 MPa, Equation (4.1) gives

$$S_e = 0.5 \, S_{ut} = 0.5 \times 1240 = 620 \text{ MPa}$$

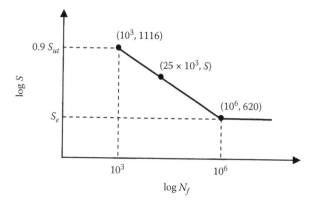

FIGURE 4.5 Schematic of the *S-N* curve for Example 3.1.

Assume a linear relationship on a log-log plot *S-N* curve between 10^3 and 10^6 cycles. The finite life of 25×10^3 cycles is between 10^3 and 10^6 cycles. Therefore, a linear relationship on the *S-N* curve of the following form applies

$$\log S = A + B \log N_f \tag{4.3}$$

where *A* and *B* are constants. Figure 4.5 shows a schematic of the *S-N* curve for this problem. Two known points on the *S-N* curve are $(10^3, 0.9\ S_{ut})$ and $(10^6, S_e)$, using which, constants *A* and *B* can be evaluated as

$$\log 0.9 S_{ut} = A + B \log 10^3$$

$$\log S_e = A + B \log 10^6 \tag{4.4}$$

Solving for *A* and *B*, we find

$$\begin{cases} A = \log \dfrac{\left(0.9 S_{ut}\right)^2}{S_e} \\[4mm] B = -\dfrac{1}{3} \log \dfrac{0.9 S_{ut}}{S_e} \end{cases} \tag{4.5}$$

Substituting for $S_{ut} = 1240$ MPa and $S_e = 620$ MPa in Equation (4.5) yields

$$A = 3.303 \text{ and } B = -0.085$$

Having determined *A* and *B*, Equation (4.3) can be written in the form:

$$S = 10^A\ N_f{}^B = 2009 \times N_f{}^{-0.085} \tag{4.6}$$

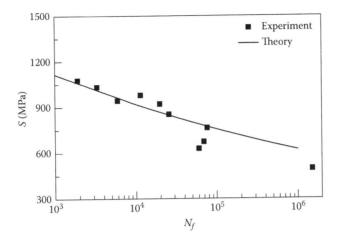

FIGURE 4.6 Comparison of experimental results and analysis for the *S-N* curve of Steel 4330 undergoing rotating-bending fatigue test.

Therefore, the stress amplitude, which results in 25×10^3 life cycles can be evaluated from Equation (4.6):

$$S = 2009 \times (25 \times 10^3)^{-0.085} = 850 \text{ MPa} \qquad (4.7)$$

Let us compare the analytical equation in (4.6) with the experimental results given in Figure 4.4. Figure 4.6 shows the comparison. It can be seen that the obtained analytical *S-N* curve matches quite well with the test results. The experimental stress level for a life of 25×10^3 cycles is $S = 853$ MPa, which is close to the value obtained by Equation (4.7). It is to be noted that in Figure 4.6, for the sake of clarity, the abscissa is plotted in log scale, but not the ordinate. ▲

It is worth noting that in the preceding discussion, the effect of mean stress S_m on the life cycle of the component is not taken into account since the loading is assumed to be completely reversed with zero mean stress, $S_m = 0$. However, in practice the mean stress can be nonzero. We will now briefly discuss the effect of mean stress on the analysis of fatigue life based on some well-known models. Let us first review some basic definitions and their relationships:

$$\text{Stress amplitude:} \quad S_a = \frac{S_{max} - S_{min}}{2}$$

$$\text{Mean stress:} \quad S_m = \frac{S_{max} + S_{min}}{2} \qquad (4.8)$$

$$\text{Stress range:} \quad S = S_{max} - S_{min}$$

$$\text{Stress ratio:} \quad R = \frac{S_{min}}{S_{max}}$$

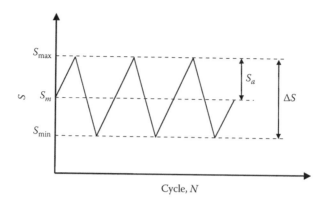

FIGURE 4.7 Schematic of alternating fatigue loading with mean value, S_m.

These definitions are schematically shown in Figure 4.7. Using these definitions, the stress ratio for a fully-reversed loading is $R = -1$.

4.2.2 EFFECT OF MEAN STRESS

During the past decades, several empirical fatigue models have been developed that take into consideration the effect of mean stress. Table 4.1 outlines the applications of some of the most popular models such as Goodman (1919); Gerber (1874); Soderberg (1939); Smith, Watson, and Topper (1970); Morrow (1965); and Walker (1970). Kujawski and Ellyin (1995) proposed a "unified approach" formulated on a fatigue damage parameter that uses the strain energy range ΔW and accounts for mean stress S_m or stress ratio R. Detailed discussion of these models is beyond the scope of this book. The interested reader is referred to publications by Dowling (2004), Kujawski and Ellyin (1995), Manson and Halford (2006), Stephens et al. (2000), Bannantine, Comer, and Handrock (1997) and Shigley and Mitchell (1983).

The following example illustrates the use of Goodman model, which relates the mean stress S_m, stress amplitude S_a, and ultimate strength S_{ut} as

$$\frac{S_m}{S_{ut}} + \frac{S_a}{S} = 1 \tag{4.9}$$

In Equation (4.9), S is the stress level for fully reversed fatigue load, $R = -1$, as given by Equation (4.3). The usefulness of Equation (4.9) is demonstrated in the following example.

Example 4.2

Suppose the axel shaft of Example 4.1 is subjected to alternating stress with non-zero mean value.

(a) Using the Goodman model, determine the fatigue life of the component if it is subjected to maximum and minimum stresses of 900 MPa and 100 MPa, respectively.
(b) Determine the allowable maximum stress for the design life of $N_f = 10^6$ with the stress ratio of $R = 0.5$.

TABLE 4.1
Some of the Well-Known Approaches to the Effect of Mean Stress

Author (Year)	Equation	Applicability
Gerber (1874)	$\dfrac{S_a}{S} + \left(\dfrac{S_m}{S_{ut}}\right)^2 = 1$	It is simple to use and works better than the Goodman model for ductile materials and low stresses.[a] It is inaccurate for compressive mean stress and notched components under tensile mean stress.[b]
Goodman (1919)	$\dfrac{S_a}{S} + \dfrac{S_m}{S_{ut}} = 1$	It is simple to use and yields conservative results for Ti-alloys. It works better than the Gerber model for high strength and low-ductility materials.[a] It can be conveniently used for compressive mean stress.[b] For tensile mean stress, it is excessively conservative and inaccurate.[c]
Soderberg (1939), S_y is the yield strength	$\dfrac{S_a}{S} + \dfrac{S_m}{S_y} = 1$	It is recommended for use in the absence of any other guidelines.[d] In many cases it overestimates the effect of mean stress.[e] It is not a valid criterion for NiTi material.[f]
Morrow (1965), σ_f is the fatigue strength coefficient[g]	$\dfrac{S_a}{S} + \dfrac{S_m}{\sigma_f} = 1$	It can be conveniently used for compressive mean stress.[b] It works well for steels, but not for aluminum alloys. An alternative form with $\tilde{\sigma}_{fB}$ (true fracture strength) instead of σ_f works quite well.[c]
Smith–Watson–Topper (1970)	$\dfrac{S_a}{S} = \left(\dfrac{1-R}{2}\right)^{0.5}$	It is a convenient choice for general use.[c]
Walker (1970), γ is called the fitting constant	$\dfrac{S_a}{S} = \left(\dfrac{1-R}{2}\right)^{1-\gamma}$	It is an excellent choice for general use; however, the value of γ is not available for a variety of materials.[c]

[a] From Schijve, J., *Fatigue of Structures and Materials*, Kluwer Academic Publishers, Dordrecht, The Netherlands, 2001.

[b] From Stephens, R.I., Fatemi, A., Stephens, R.R. and Fuchs, H.O., *Metal Fatigue in Engineering,* 2nd ed., John Wiley & Sons, Inc., New York, 2000.

[c] From Dowling, N.E., Mean Stress Effects in Stress-Life and Strain-Life Fatigue, SAE Fatigue 2004, *Proc. 3rd Int. SAE Fatigue Congress*, Sao Paulo, Brazil, 2004.

[d] From Schwarts, R.T., Correlation of Data on the Effect of Range of Stress on the Fatigue Strength of Metals for Tensile Mean Stresses, Master's thesis, Ohio State University, Columbus, OH, 1948.

[e] From Woodward, A.R., Gunn, K.W. and Forrest, G., *Int. Conf. Fatigue Metals*, Institution of Mechanical Engineers, Westminster, London, 1156–1158, 1956.

[f] From Tabanli, R.M., Simha, N.K. and Berg, B.T., *Mater. Sci. Eng.* A 273–275, 644–648, 1999.

[g] Alternative form is also used by substituting σ_f for the true fracture strength $\tilde{\sigma}_{fB}$. However, the value of $\tilde{\sigma}_{fB}$ is not available for a variety of materials.

Solution

(a) First, let us determine the stress amplitude and mean stress from Equation (4.8):

$$S_a = \frac{S_{max} - S_{min}}{2} = \frac{900 - 100}{2} = 400 \quad \text{MPa}$$

$$S_m = \frac{S_{max} + S_{min}}{2} = \frac{900 + 100}{2} = 500 \quad \text{MPa}$$

$$R = \frac{100}{900} = 0.1$$

Using the Goodman model in Equation (4.9), the stress amplitude, S, can be determined as follows:

$$S = \frac{S_a S_{ut}}{S_{ut} - S_m} = \frac{400 \times 1240}{1240 - 500} = 670 \quad \text{MPa}$$

Having determined the stress level S, we use Equation (4.6) to evaluate the number of cycles to failure as

$$S = 670 = 2009 \times (N_f)^{-0.085}$$

$$N_f = \left(\frac{2009}{670}\right)^{1/0.085} = 4 \times 10^5 \quad \text{cycles}$$

(b) Since the component is designed for infinite life ($N_f = 10^6$), we simply substitute S_e for S in Equation (4.9):

$$\frac{S_m}{S_{ut}} + \frac{S_a}{S_e} = 1 \tag{4.10}$$

$$\frac{S_m}{1240} + \frac{S_a}{620} = 1 \tag{4.11}$$

There are two unknowns in Equation (4.11), that is, S_a and S_m. Another relationship between S_a and S_m in terms of the load ratio can be derived as follows:

$$\frac{S_a}{S_m} = \frac{S_{max} - S_{min}}{S_{max} + S_{min}} = \frac{1 - S_{min}/S_{max}}{1 + S_{min}/S_{max}} = \frac{1 - R}{1 + R}$$

$$\frac{S_a}{S_m} = \frac{1 - 0.5}{1 + 0.5} = \frac{1}{3} \tag{4.12}$$

Equation (4.11) along with Equation (4.12) can be solved to find S_a and S_m:

$$S_a = 248 \text{ MPa}$$

$$S_m = 744 \text{ MPa}$$

Having determined S_a and S_m, the maximum allowable stress S_{max} can now be evaluated from Equation (3.8):
$$S_{max} = S_a + S_m = 248 + 744 = 992 \text{ MPa} \blacktriangle$$

4.2.3 LOAD HISTORY

Traditionally, most fatigue data are presented for constant stress amplitude under sinusoidal loading as represented by the *S-N* curve. However, machine elements in practical applications experience a complex spectrum of loading histories with variable amplitudes and frequencies. For example, a windmill blade experiences a centrifugal dead weight load in addition to a very complex combination of low- and high-wind loading amplitudes during its normal operation. Figure 4.8 shows a schematic of the variable load history. It shows that not only can the mean stress change, but also the stress amplitude can be variable. Indeed, the load spectrum can be much more complex involving a random load-cycle pattern in the case of, for example, an aircraft during operation from taxi to takeoff to landing (Min, Xiaofei, and Qing-Xiong 1995). More comprehensive review can be found in Newmark (1952), Kaechele (1963), Fatemi and Yang (1998), and more recently, Pook (2007).

Determination of the allowable stress requires data on the load spectra of the fatigue crack growth behavior (Fingers et al. 1980). Early attempts to study fatigue life of components undergoing variable stress amplitude date back to pioneering works of Palmgren (1924), Langer (1937), and Miner (1945). Palmgren first proposed the linear rule in connection to his work on ball bearing fatigue life estimation. This method was later extended by Miner account for the load-cycle variation with application to aircraft design. This linear damage

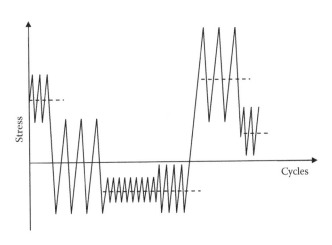

FIGURE 4.8 Variable load history.

rule basically accounts for cumulative damage induced at each stress amplitude S_i. It states that the damage fraction, D_i, at stress level S_i is equal to the cycle ratio, $N_i/N_{f,i}$, where N_i and $N_{f,i}$ are the accumulated number of cycles and the fatigue life at stress level of S_i, respectively. The fatigue criterion for variable-amplitude loading is defined as

$$\sum \frac{n_i}{N_i} = 1 \tag{4.13}$$

and assumes that the life to failure can be estimated by summing the percentage of life used up in each stress level. The linear damage rule has two shortcomings: (1) It does not consider the well-known loading *sequence* effect. Research shows that the order in which loading cycles are applied (e.g., high load followed by low load in contrast to low load followed by high load) has a significant effect on the stress-strain response of a material due to the nonlinear relationship between stress and strain. (2) The linear damage rule is amplitude independent and its prediction of damage accumulation is independent of the stress level. Indeed, experimental observations reveal that at high-strain amplitudes, a crack(s) initiates after a few cycles; whereas at low-strain amplitudes, almost all life is spent in the crack initiation phase. Despite these limitations, the use of Miner's rule is widespread due to its simplicity. Example 4.3 shows an implementation of Miner's rule in a fatigue problem with a variable-loading amplitude.

Example 4.3

Consider the axel shaft of Example 4.1. If the shaft is subjected to completely reversed loading at 700 MPa for 25000 cycles and then at 800 MPa for 15000 cycles, what would be its remaining life if it is subjected to the stress level of 900 MPa?

SOLUTION

The first step is to determine the number of cycles to failure at each stress level based on the specified information. Invoking Equation (4.6), the number of cycles to failure can be determined as

$$N_{f,i} = \left(\frac{2009}{S_i} \right)^{1/0.085} \quad \text{cycles}$$

The evaluated $N_{f,i}$ for each stress level is given in Table 4.2.

TABLE 4.2

Fatigue Life at Each Stress Level

Stress level, MPa	$S_1 = 700$	$S_2 = 800$	$S_3 = 900$
Fatigue life, cycle	$N_{f,1} = 250{,}000$	$N_{f,2} = 50{,}000$	$N_{f,3} = 12{,}000$

Using Miner's rule given by Equation (4.13), the sum of the damage at different stress level should reach 100%:

$$\frac{25000}{250000} + \frac{15000}{50000} + \frac{N_3}{12000} = 1 \qquad (4.14)$$

The estimated remaining life at stress level S_3 is

$$N_3 = 0.6 \times 12000 = 7,200 \text{ cycles} \ \blacktriangle$$

4.2.4 STRESS-STATE: TORSION, TENSION–COMPRESSION, BENDING, AND COMBINED MODE

In many engineering applications, mechanical components—crank shafts, propeller shafts, rear axles, and the like—are subjected to complex states of stress and strain in which the three principle stresses are nonproportional and/or their directions change during a loading cycle. Applied stresses on the components can vary from uniaxial to biaxial to combined combinations referred to as *multiaxial*. Multiaxial fatigue is quite complex but necessary for reliable operation of many engineering components (Bannantine, Comer, and Handrock 1989).

During the past decades, numerous models have been developed taking into account the effect of multiaxiality on fatigue life. These models are categorized as stress-based, strain-based, and energy-based criteria (Karolczuk and Macha 2005). Some reviews of multiaxial fatigue models are available in the works by Garud (1981), You and Lee (1996), Papadopoulos et al. (1997), Backstrom and Marquis (2001), Wang and Yao (2004), and Karolczuk and Macha (2005). A comprehensive report of multiaxial fatigue analyses is also available in a book by Socie and Marquis (2000). Among the available multiaxial models, the so-called critical plane approach proposed first by Stanfield (1935) is popular. This model assumes that the crack initiates and propagates on a so-called critical plane across the material (Socie 1993). Accordingly, fatigue life analysis is performed on the critical plane on which the amplitude of shear stress (strain) or a combination of shear and normal stress (strain) attains its maximum value.

A simpler approach uses the concept of equivalent stress, which assumes that the multiaxial stress components can be converted into a single equivalent stress, S_{eq}, so that the problem can be treated as a uniaxial fatigue problem. For example, for a two-dimensional stress state, the Von Mises equivalent stress amplitude S_{eq} is given as (Stephens et al. 2000)

$$S_{eq} = \sqrt{S_x^2 + S_y^2 - S_x^2 S_y^2 + 3\tau_{xy}^2} \qquad (4.15)$$

where S_x, S_y, and τ_{xy} are normal stress in the x direction, normal stress in the y direction, and shear stress on xy plane, respectively (Figure 4.9).

Note that Equation (4.15) is applicable to completely reversed loading with zero mean stress. The following example illustrates how this method is applied.

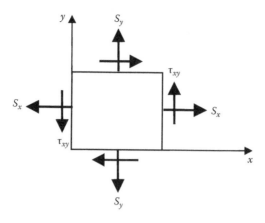

FIGURE 4.9 Stress element in multiaxial loading.

Example 4.4

Consider a shaft of $D = 15$ mm diameter is made of the same material as in Example 4.1. The shaft is subjected to completely reversed torque of $T = 250$ Nm and axial load of $F = 50$ kN. What is the expected fatigue life considering the effective stress method?

Solution

The first step is to determine normal and shear stresses on the surface of the shaft:

$$S_x = \frac{4F}{\pi D^2} = \frac{4 \times 50 \times 10^3}{\pi (0.015)^2} = 283 \quad \text{MPa}$$

$$S_y = 0$$

$$\tau_{xy} = \frac{16T}{\pi D^3} = \frac{16 \times 250}{\pi (0.015)^3} = 377 \quad \text{MPa}$$

The alternating effective stress is calculated from Equation (4.15):

$$S_{eq} = \sqrt{(283)^2 + 3(377)^2} = 711 \quad \text{MPa}$$

Having determined the stress level S_{eq}, we use Equation (4.6) to evaluate the number of cycles to failure as

$$S_{eq} = 711 = 2009 \times (N_f)^{-0.085}$$

$$N_f = \left(\frac{2009}{711}\right)^{1/0.085} = 2 \times 10^5 \text{ cycles } \blacktriangle$$

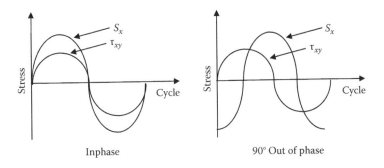

FIGURE 4.10 Proportional and nonproportional multiaxial loading.

The method of equivalent stress amplitude is a simple and convenient way of determining fatigue life subjected to proportional multiaxial loading. Proportionality refers to in-phase loading during which different stresses reach their peaks simultaneously. In contrast, in nonproportional loading, stresses are out of phase (Figure 4.10). The effective stress method falls short in addressing nonproportionality, and thus, one must invoke other multiaxial models such as the critical plane model.

4.3 FATIGUE AND ENERGY DISSIPATION

The mechanism of energy transformation and its relationship to fatigue life is the subject of this section. First, a brief discussion of micro/nanoscale mechanisms of energy dissipation during deformation is presented. Next, it is shown how plastic energy dissipation is related to fatigue. The irreversibilities associated with the energy conversion are presented and discussed in terms of entropy in Chapters 5 and 6. However, temperature as the thermodynamic conjugate variable of the entropy and its role in fatigue analysis are the main focus of Section 4.4.

4.3.1 MICRO/NANOSCALE MECHANISM OF ENERGY DISSIPATION

The physical underpinning of the notion of dislocation as a useful tool capable of addressing the dissipation mechanism at atomic scale has been extensively investigated for many decades. It is widely accepted that the dissipative feature of materials during deformation can be attributed to dislocation. According to Ashby and Jones (2012): "dislocations are the carriers of deformation, much as electrons are the carriers of charge." The origin of this notion goes back to the pioneering work of Eshelby (1949a,b) who applied the concept of dislocation to explain the mechanism of the energy dissipation in metals for the first time. Using the theory of internal friction introduced by Zener (1937, 1938, 1940, 1948), Eshelby went on to describe the mechanism of mechanical damping in vibrating metals by taking into consideration the role of moving dislocations on the damping. Eshelby postulated that in addition to the thermoelastic damping (introduced by Zener), there is an additional energy loss due to the oscillation of dislocations. The motion of dislocations leads to redistribution of stresses in the material which, in turn, results in change in the temperature distribution at any locations. The resulting temperature gradient drives the heat flow and causes mechanical damping. This damping depends on the applied stress and the density of dislocations. Eshelby (1949a) showed that the mechanical damping Q^{-1} due to the harmonic oscillation of dislocations is given by

$$Q^{-1} = \frac{2\pi}{10\,\alpha}\frac{c_p - c_v}{c_p}G^2 a^2 fn \log\left(\frac{\alpha}{2\pi f l^2}\right)\left(\frac{d}{\sigma_0}\right)^2 \tag{4.16}$$

where c_p, c_v, and α are the specific heat capacity at constant pressure, specific heat capacity at constant volume, and thermal diffusivity, respectively. The parameters f and d represent the frequency and amplitude of the oscillation, G is the shear modulus, a denotes the interatomic spacing, n is the number of dislocations per unit area, and σ_0 is the shear stress amplitude. The parameter l is the so-called cut-off length which represents the distance from the dislocation below which the elastic solution of the problem is not valid. Following the work of Lawson (1941), Eshelby presented a numerical example of energy loss for polycrystalline specimens of very pure copper subjected to the compressive stresses on the order of 1000 kPa at the frequency of 50 kHz. He showed that damping doubled from $Q^{-1} = 2 \times 10^{-3}$ before compression to $Q^{-1} = 4 \times 10^{-3}$ after compression. Eshelby (1949) pointed out that while the model of energy dissipation of oscillating dislocations provides an acceptable order of magnitude, it falls short of providing evidence for the effect of frequency on the damping.

A theoretical study by Mura and Lyons in 1966 improved Eshelby's model of energy dissipation by considering a continuous distribution of dislocations instead of discrete dislocations. The result of their work renders a model for evaluation of damping in metals taking into account the effect of frequency. The damping is given as

$$Q^{-1} = \frac{\Omega M}{(1-\Omega^2)^2 + \Omega^2[M^2 + N(1-\Omega^2)]^2} \tag{4.17}$$

where Ω is proportional to the frequency f and is defined, along with other parameters as

$$\Omega = 2\pi f\left(\rho/(\lambda+2\mu)c^2\right)^{0.5}$$
$$M = \left[\rho c^2/(\lambda+2\mu)^{0.5}\right]\left[(3\lambda+2\mu)^2\alpha_T^2 T_0/k(\lambda+2\mu)\right] \tag{4.18}$$
$$N = \left[(\lambda+2\mu)/\rho c^2\right]^{0.5}(\rho c_v/k)$$

where ρ is the density, λ and μ are Lamé constants, α_T represents the linear thermal expansion coefficient, k denotes the thermal conductivity, T_0 is the reference temperature, and $(1/c)$ is the dislocation-wall distance or the grain size. A numerical example of the quantity of damping Q^{-1} against Ω is presented in Figure 4.11. The selected material is copper with the following properties: $\rho = 8900$ kg/m³, $c_v = 385.6$ J/(kg·K), $k = 418.7$ (W/m·K), $\alpha = 17.71 \times 10^{-6}$ (1/°C), $\lambda = 9.5 \times 10^{10}$ (N/m²), and $\mu = 4.5 \times 10^{10}$ (N/m²). Different values of $(1/c)$ are considered in this figure. Figure 4.11 shows that there is a peak in energy loss at $\Omega = 3 \times 10^{-n}$ which cannot be predicted by Eshelby's model. Note that here n is defined as the power index.

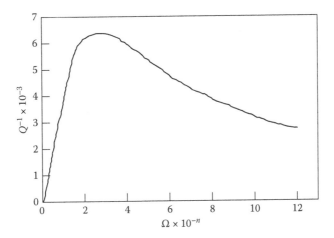

FIGURE 4.11 Damping Q^{-1} in copper as a function of frequency Ω for different values of $n =$ 5,4,3. (Adopted from Mura, T. and Lyons, W.C., *J. Acoust. Soc. Am.* 39, 527–531, 1966.)

4.3.2 MACROSCALE MECHANISM OF ENERGY DISSIPATION

Let us turn our attention to the macroscale (bulk) manifestation of energy dissipation in relation to fatigue life assessment. Fatigue failure of metals can be well described and predicted through the analysis of the strain energy dissipated as hysteresis heat. The service life of a metallic component subjected to fatigue load is directly related to crack initiation and propagation periods. During each period, dissipation of energy associated with cyclic deformation manifests itself in the form of heat generation (Allen 1985; Blotny et al. 1986). For deformations at low stress levels, elastic strain energy is dominant and the process between the external work and the strain energy is reversible. As discussed in Section 4.2, these processes are categorized as high-cycle fatigue, and their associated heat generation is often quite small. In contract, as the material is cyclically loaded beyond it elastic limit, the mechanical input work is expended in plastically deforming the material, converting it into heat that raises the temperature of the body and transfers to the surroundings.

4.3.3 PREDICTION OF FATIGUE FAILURE BASED ON ENERGY DISSIPATION

The application of energy approach for the prediction of fatigue has a rich history with voluminous supporting literature, for example, Feltner and Morrow (1961); Morrow (1965); Kaleta, Blotny, and Harig (1990); Golos (1995); Glinka, Shen, and Plumtree (1995); Park and Nelson (2000); Meneghetti (2006); and Charkaluk and Constantinescu (2009). In particular, Morrow's work (1965) is largely considered to be one of the pioneering works on the assessment of energy generation during fatigue loading. In his model, plastic strain energy is evaluated for fatigue tests from cyclic stress-strain properties, and a hysteresis loop is utilized for the assessment of energy dissipation per cycle. The strain energy dissipation per cycle w_p is then expressed as

$$w_p = \frac{4\sigma'_f\left(\frac{1-n'}{1+n'}\right)}{\left(\varepsilon'_f\right)^{n'}}\left(\frac{\varepsilon_p}{2}\right)^{1+n'} \tag{4.19}$$

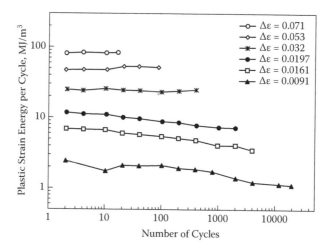

FIGURE 4.12 Plastic strain energy dissipation per cycle as a function of number of cycles. (Reproduced from Morrow, J.D., Cyclic Plastic Strain Energy and Fatigue of Metals, *Internal Friction, Damping, and Cyclic Plasticity, ASTM STP* 378, Philadelphia, PA, 45–84, 1965.)

where $\Delta\varepsilon_p$ is the plastic strain range, n′ is the cyclic strain hardening exponent, and ε'_f and σ'_f are cyclic ductility and fatigue strength coefficient, respectively. Morrow (1965) experimentally demonstrated that in fully reversed fatigue tests, the amount of energy generation per cycle is approximately constant, but varies with the strain level, $\Delta\varepsilon$, and the cyclic properties of the material. Figure 4.12 (reproduced from Morrow 1961) shows the plastic strain energy per cycle for a nickel-base alloy as a function of number of cycles. It can be seen that the plastic energy dissipation is approximately constant throughout the low-cycle fatigue test. This finding enables one to correlate the amount of plastic energy dissipation with the fatigue life of the samples tested.

Equation (4.19) can be rewritten in the following useful form to relate fatigue life, N_f, and the energy dissipation per cycle, w_p:

$$w_p = 4\varepsilon'_f\,\sigma'_f\left(\frac{c-b}{c+b}\right)\left(N_f\right)^{b+c} \tag{4.20}$$

where b and c are the fatigue strength exponent and the fatigue ductility exponent, respectively.

Morrow's finding that the dissipation of plastic strain energy per cycle is nearly constant implies that the plastic strain energy accumulates linearly during the fatigue life. It follows immediately that the total accumulation of the plastic energy at each step up to N number of cycles is

$$W_p^N = \sum_1^N\ w_p = N\ \ w_p \tag{4.21}$$

To take into account the contribution of the elastic energy W_e, Park and Nelson (2000) introduced an additional term and combined the two energy terms to obtain the total strain energy W_t as follows:

$$W_t = W_p + W_e = AN_f^\alpha + BN_f^\beta \tag{4.22}$$

where constants A, α, B, and β can be determined from a set of uniaxial fatigue test data. The total strain energy encompasses a wide range of fatigue cycles, from low to high cycles. It is generally accepted that in the low-cycle regime, the plastic strain energy W_p is dominant while the elastic strain energy W_e dominates the high-cycle fatigue.

As discussed by Park and Nelson (2000), the effect of stress state on plastic strain energy can be effectively taken into account by introducing the so-called multiaxiality factor (MF). Effective plastic work per cycle Δw^* is then defined by multiplying the MF by the plastic work per cycle, that is, $\Delta w^* = MF\Delta w$. The effective plastic strain energy Δw^*, instead of plastic strain energy Δw, can be utilized to account for the effect of stress state on entropy generation.

The relationship between the energy dissipation and fatigue life and its implication on material degradation remains to be of interest to many researchers (e.g., Halford 1966; Kaleta et al. 1990; Skelton 2004). The beauty of applying the principle of energy generation to fatigue processes is that the combined effect of many factors—stress state, stress level, environmental conditions, and so forth—on the accumulation of fatigue can be accounted for. Clearly, the dissipation energy manifests itself as the temperature rise of the specimen. Therefore, an immediate question of whether fatigue life can be related to the rise of specimen temperature is taken up in the next section.

4.4 FATIGUE–TEMPERATURE RISE

Dislocation movement in metals is the main cause of heat dissipation and consequently, rise in temperature (Hirth and Lothe 1968; Kuhlmann-Wilsdorf 1987; Moore and Kuhlman-Wilsdorf 1970; Nabarro 1967; Stroh 1953). In fact, for a wide range of metals subjected to deformation at moderate to high strain rate, almost all the plastic strain energy is converted into heat (Kapoor and Nemat-Nasser 1998; Padilla et al. 2007). Specifically, Hodowany (1997) states that over a wide range of strain rates, almost 90% of the input plastic work converts into heat. It should be pointed out that for some metals with hexagonal close packed (hcp) structure such as zirconium and hafnium, a considerable fraction of plastic energy is not dissipated as heat, but stored in the material (Padilla et al. 2007; Subhash, Ravichandran, and Pletka 1997). If the rate of heat generation is greater than the dissipation to the surroundings, the temperature of the specimen must necessarily increase.

4.4.1 Temperature Evolution during Fatigue

Figure 4.13 illustrates the effect of elastic and plastic strain energies on the variation of the temperature of a body subjected to fatigue load. Plastic strain and its associated dissipation

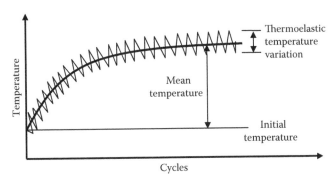

FIGURE 4.13   Idealized plot of temperature variation during a fatigue test.

result in an increase in the average temperature while the material response to elastic strain is simply a cyclic change in temperature superimposed on the plastic response. In a low-cycle regime, where the plastic deformation is dominant, the average temperature rise is significantly higher than the cyclic temperature variation contributed by the elastic strain. Wang et al. (2000), for example, showed that fatigue tests at 1000 Hz could raise the average temperature from 200 to 400 K above the initial temperature, depending on the material tested and the specimen geometry. This rise in temperature is a useful probe for several important purposes such as measuring crack propagation (Blotny et al. 1986), evaluating the energy required to produce a unit area of a fatigue crack by propagation (Gross 1981), determining the endurance limit of some materials (Luong 1998), and characterizing the evolution of cumulative damage in fatigue processes (Amiri, Naderi, and Khonsari 2011).

These temperature variations can be easily detected by infrared (IR) thermography technique as a function of time. Both theoretical and experimental methods are developed for determination of temperature rise and the assessment of related fatigue life. Some of the pertinent research has been published recently by Meneghetti (2006); Ranc, Wagner, and Paris (2008); Amiri and Khonsari (2010a, 2010b); and Amiri, Naderi, and Khonsari (2011).

State-of-the-art IR thermography has made it is possible to capture very small variation in temperature due to elastic deformation which was somewhat impractical when dealing with traditional sensors such as thermocouples. A review of the available methods and technologies for fatigue failure prediction based on temperature evolution is presented by Amiri and Khonsari (2010b). This reference reports the results of an extensive set of bending fatigue tests with different metals, and the actual temperature evolution of specimens was recorded and analyzed. There, it was shown that fatigue life can be ideally characterized in three distinct phases as illustrated in Figure 4.14. During the first phase, the temperature of the specimen increases rapidly from its initial temperature. The second phase begins when the temperature levels off and becomes steady, and the third phase is associated with the sharp increase in temperature after which fracture occurs. During the first phase, the hysteresis heat generation is greater than the heat transferring out of the specimen; hence, the temperature rises from the initial temperature T_o to T^*. During the second phase, when temperature becomes stable, there is a balance between the heat generation and heat dissipation by means of conduction, convection, and radiation mechanisms. Most of the fatigue life is

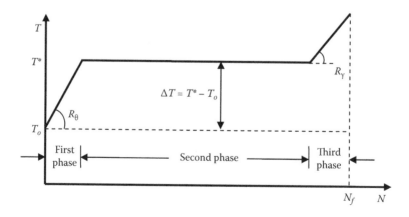

FIGURE 4.14 An idealized temperature evolution during a fatigue test.

spent in the second phase. Laboratory tests reveal that the rate of increase in temperature in the first phase, R_θ, and the stabilized temperature in second phase, T^*, increase as the load increases. The rapid temperature rise with rate of R_γ in the third phase is associated with the plastic deformation at the crack tip, and the plastic work generated during this deformation is mostly converted into heat (Morrow 1965).

In the following example, we demonstrate how temperature rise during a low-cycle fatigue process can be simply simulated by considering plastic energy dissipation as a heat source. Simulation can effectively predict the first and second phases of temperature rise, while the third phase requires information about size of the macrocrack and its propagation rate.

Example 4.5

A plate specimen with the geometry shown in Figure 4.15 undergoes axial fatigue load. If the plate is initially at room temperature, T_0, find the transient temperature evolution.

Due to symmetry of load, a symmetric temperature profile can be assumed along the length of the specimen. Assume that the temperature variation across the thickness and along the width is negligible and consider the one-dimensional thermal analysis. All the material's properties (density, ρ; specific heat capacity, c_p; and thermal conductivity, k) are assumed to be uniform and known.

SOLUTION

The system boundary and the coordinate system are shown in Figure 4.15. For one-dimensional analysis, we consider the heat conduction along the x axis. The governing equation, and the boundary and initial conditions are

$$\rho c_p \frac{\partial T}{\partial t} = k \frac{\partial^2 T}{\partial^2 x} + w, \quad -L \le x \le L, 0 < t \tag{4.23a}$$

$$T(x = L, t) = T(x = -L, t) = T_0, \ 0 < t \tag{4.23b}$$

$$T(x, t = 0) = T_0, \ -L \le x \le L \tag{4.23c}$$

where Δw (W/m³) is the volumetric energy generation inside the system. It is assumed that temperature at both ends, $x = \pm L$, is fixed at room temperature T_0:

$$\theta = T - T_0 \tag{4.24}$$

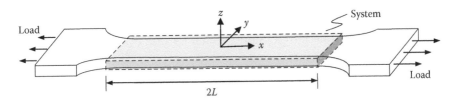

FIGURE 4.15 Geometry of the specimen under fatigue load.

Due to the symmetry of the problem, let us find the solution for half-length. Equations (4.23) read

$$\frac{1}{\alpha}\frac{\partial \theta}{\partial t} = \frac{\partial^2 \theta}{\partial^2 x} + \frac{w}{k}, \quad 0 \leq x \leq L, \ 0 < t \tag{4.25a}$$

$$\partial \theta / \partial x \ (x = 0, t) = \theta(\, x = L/2, t) = 0, \ 0 < t \tag{4.25b}$$

$$\theta(x, t = 0) = 0, \ 0 \leq x \leq L \tag{4.25c}$$

where $\alpha = k/\rho c_p$ is the thermal diffusivity. The solution of the partial differential equation in Equation (4.25a) subjected to boundary and initial conditions in Equations (4.25b) and (4.25c) can be obtained using the integral transform technique (Ozisik 2002). This technique provides a very useful tool for analysis of heat conduction problems since it has no inversion difficulties as is the case in Laplace transformation. The solution of the temperature distribution is in the form of infinite series:

$$\theta(x,t) = \sum_{m=0}^{\infty} e^{-\alpha \beta_m^2 t} K(\beta_m, x) \int_{t'=0}^{t} e^{\alpha \beta_m^2 t'} \cdot A(\beta_m, t') dt' \tag{4.26a}$$

where

$$A(\beta_m, t') = \frac{\alpha}{k} \int_{x'=0}^{L} K(\beta_m, x') \cdot w \, dx' \tag{4.26b}$$

where $K(\beta_m, x)$ is the kernel with its associated eigenvalues β_m:

$$K(\beta_m, x) = \sqrt{\frac{2}{L}} \cos(\beta_m x) \tag{4.26c}$$

$$\beta_m = \frac{(2m+1)\pi}{2L} \text{ with } m = 0, 1, 2, 3, \dots \tag{4.26d}$$

Evaluating the integral in Equation (4.26a) and simplifying the solution yields

$$\theta(x,t) = \frac{2}{kL} w \sum_{m=0}^{\infty} \frac{(-1)^m}{\beta_m^3} \cos(\beta_m x)(1 - e^{-\alpha \beta_m^2 t}) \tag{4.27}$$

Figure 4.16 shows the evolution of temperature rise θ at given location x. The θ_s denotes the steady-state temperature, which can be evaluated by setting $t \to \infty$ in Equation (4.27):

$$\theta_s(x,t) = \frac{2}{kL} w \sum_{m=0}^{\infty} \frac{(-1)^m}{\beta_m^3} \cos(\beta_m x) \tag{4.28}$$

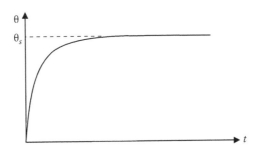

FIGURE 4.16 Typical temperature evolution during fatigue.

The one-dimensional thermal analysis gives an acceptable result for uniaxial fatigue of a plate specimen for which the axial heat conduction is the dominant mode of heat transfer. However, in problems with comparable lateral heat conduction and/or significant convection and radiation heat transfer, a 2D or 3D analysis is required. In the following example, a 2D thermal analysis presented by Amiri and Khonsari (2010a) is employed to simulate temperature variation in bending fatigue. ▲

Example 4.6

The surface temperature of plate specimens subjected to bending fatigue is shown in Figure 4.17. These results pertain to bending fatigue of Aluminum 6061-T6 at frequency of $f = 10$ Hz. The specific heat capacity, c_p; density, ρ; and thermal conductivity, k, of the material are 893 (J/kgK), 2710 (kg/m³), and 273 (W/mK), respectively. Test conditions are given in Table 4.3. The specimen is clamped at one end and the other end oscillates at a displacement amplitude of δ (see Figure 4.18).

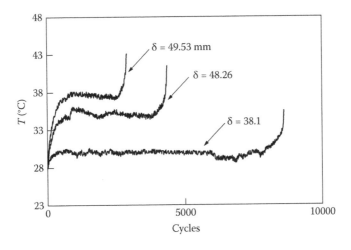

FIGURE 4.17 Temperature evolutions for three different displacement amplitudes. (Reproduced from Amiri, M. and Khonsari, M.M., *Int. J. Fatigue* 32, 382–389, 2010a.)

TABLE 4.3

Test Conditions

δ (mm)	h (W/m²K)	$g \times 10^3$	T_0 (°C)
38.1	120	15.4	28
48.26	120	84	28
49.53	120	101.5	28

The clamped end is insulated to reduce heat transfer to the grip. Assuming the convective heat transfer is dominant, employ the 2D model for thermal analysis to determine the temperature evolution and compare the results with the experimental work of Amiri and Khonsari (2010a).

SOLUTION

The specimen is modeled as a rectangle with designated coordinate system shown in Figure 4.19. The free end is assumed to be at room temperature. The top and bottom surfaces exchange heat via convection with the surroundings. The heat generation, Δw, is assumed to be uniformly distributed in the region. Similar to Equations (4.23), the governing equation in 2D and the initial and boundary conditions can be written as

$$\frac{1}{\alpha}\frac{\partial T}{\partial t} = \frac{\partial^2 T}{\partial^2 x} + \frac{\partial^2 T}{\partial^2 y} + \frac{w}{k}, \ 0 \le x \le L, \ 0 \le y \le a, \ 0 < t \qquad (4.29a)$$

$$T(0, y, t) = T_0, \ \partial T/\partial x(L, y, t) = 0, \quad 0 < t \qquad (4.29b)$$

$$-k\partial T/\partial y(x, 0, t) + h(T-T_0) = 0, \ k\partial T/\partial y(x, a, t) + h(T-T_0) = 0, \ 0 < t \quad (4.29c)$$

$$T(x, y, 0) = T_0, \ 0 \le x \le L, \quad 0 \le y \le a \qquad (4.29d)$$

where h is the convection heat transfer coefficient given in Table 4.2. Using the integral transform method explained in Example 4.5, the solution to the problem in Equations (4.29) is given as (Amiri and Khonsari 2010a)

$$T(x,y,t) = (\theta + 1)T_0 \qquad (4.30)$$

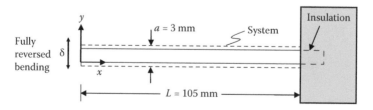

FIGURE 4.18 Specimen undergoing bending fatigue load.

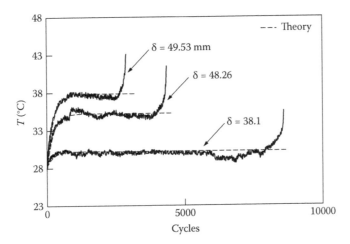

FIGURE 4.19 Temperature evolutions for three different displacement amplitudes as compared with theory.

where

$$\theta(\xi, \eta, \tau) = \sum_{m=1}^{\infty} \sum_{n=1}^{\infty} \frac{1}{\beta_m^2 + \lambda_n^2} \left[1 - e^{-\alpha\left(\beta_m^2 + \lambda_n^2\right)\tau} \right] \cdot \left[\left(\lambda_n^2 + B^2\right)\left(\frac{a}{L} + \frac{B}{\lambda_n^2 + B^2}\right) + B \right]^{-1} \cdot \sin\beta_m \xi$$

$$\cdot \left(\lambda_n \cdot \cos\lambda_n \eta + B \cdot \sin\lambda_n \eta\right) \int_{\xi'=0}^{1} \int_{\eta'=0}^{a/L} \sin\beta_m \xi' \cdot \left(v_n \cdot \cos\lambda_n \eta' + B \cdot \sin\lambda_n \eta'\right) \cdot g(\xi', \eta') \cdot d\xi' d\eta'$$

(4.31)

where β_m, and λ_n are eigenvalues for x and y directions, respectively. The dimensionless parameters θ, ξ, η, τ, B, and g are defined as follows:

$$\theta = \frac{T - T_0}{T_0}$$

$$\xi = \frac{x}{L}$$

$$\eta = \frac{y}{a}$$

$$\tau = \frac{\alpha t}{L^2}$$

$$B = \frac{hL}{k}$$

$$g = \frac{f \, wL^2}{kT_0}$$

Substituting the material properties and testing conditions reported in Table 4.3, into Equation (4.31) for three different displacement amplitudes results in the temperature variation as plotted in Figure 4.19. Note that experimental results are also presented in this figure for comparison. ▲

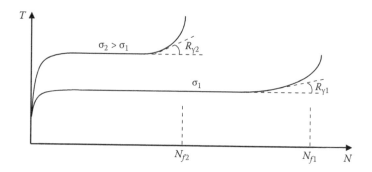

FIGURE 4.20 Slope of temperature rise in third phase at two stress levels.

4.4.2 APPLICATION TO PREDICTION OF FATIGUE FAILURE

The third phase of temperature evolution can be utilized as an immediate warning of imminent failure. Huang et al. (1984) showed that for materials with high ductility, the slope of the temperature rise at the beginning of the third phase, R_γ, can be correlated to the fatigue life, N_f (see Figure 4.20). The correlation is given as

$$R_\gamma = C' \cdot \exp\left(\frac{G}{\left(N_f\right)^{1/b}} \right)$$ (4.32)

where C', G, and b are constants that depend on the properties of the material and the test conditions. The methodology presented by Huang et al. may be useful for terminating the operation of machinery in order to prevent catastrophic failure.

Jiang et al. (2001) showed that the life of specimen, N_f, undergoing axial fatigue load is related to the temperature rise from the initial temperature, $\Delta T = T^* - T_0$ (see Figure 4.21) and presented the following correlation:

$$\left(N_f\right)^m = C \quad T$$ (4.33)

where m and C are material constants. Jiang et al. suggested that the temperature rise during the second phase can be used as an index of fatigue life.

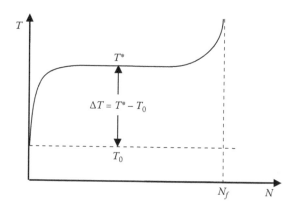

FIGURE 4.21 Steady-state temperature rise in second phase.

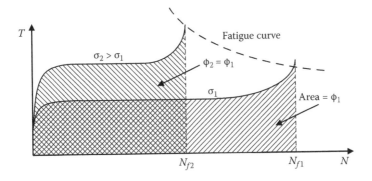

FIGURE 4.22 Area under the temperature profile at two stress levels.

Fargione et al. (2002) have proposed another fatigue life prediction method for uniaxial loading based on the area under the temperature rise over the entire number of cycles. Shown in Figure 4.22 is the area underneath the temperature profile for two different stress amplitudes, σ_1 and σ_2, denoted by ϕ_1 and ϕ_2, respectively. Experimental work of Fargione et al. revealed that the area ϕ is a constant regardless of the stress amplitude. It can be expressed as

$$\phi = \int_0^{N_f} T\, dN = \text{constant} \tag{4.34}$$

Having evaluated ϕ for one test, the entire fatigue curve can be obtained. In their study, the fatigue curve is defined as the locus of the maximum values of the temperature at the end of the test (see Figure 4.22).

An interesting technique for assessment of fatigue life is presented by Meneghetti (2007), which involves measuring the cooling rate after a sudden interruption of the fatigue test (see Figure 4.23). After the steady-state temperature, T^*, is attained, the test is stopped to measure the slope of cooling curve, dT/dt. Upon interruption of the test, the rate of cooling is correlated to the specific heat energy, Q, via

$$Q = -\frac{\rho c}{f} \frac{dT}{dt}\bigg|_{t=t^*} \tag{4.35}$$

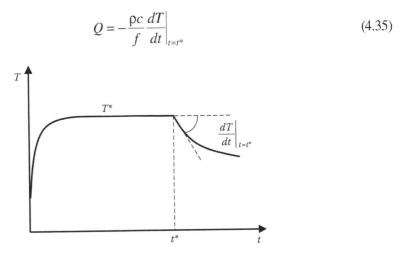

FIGURE 4.23 Slope of the cooling rate after stopping the test. (From Meneghetti, G., *Int. J. Fatigue* 29, 81–94, 2007.)

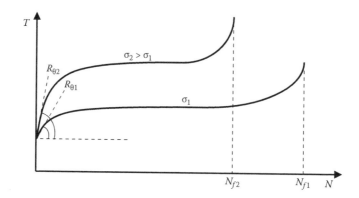

FIGURE 4.24 Slope of temperature at the beginning of the test at two stress levels.

where ρ, c, and f are density, specific heat capacity, and frequency, respectively. Meneghetti showed that fatigue life is correlated to the specific heat energy through the following expression:

$$N_f = C\left(Q\right)^m \tag{4.36}$$

where C and m are material constants. Having determined Q from the slope of the cooling curve, the fatigue life can, then, be estimated. This technique, also, provides a quick estimation of the energy that converts into heat during the fatigue process.

Amiri and Khonsari (2010a, 2010b) developed a method for prediction of fatigue failure, N_f, based on the initial slope of temperature rise, R_θ, as shown in Figure 4.24. Their results showed that there exists a relationship between N_f and R_θ as

$$N_f = c_1 R_\theta^{c_2} \tag{4.37}$$

where c_1 and c_2 are constants. Their experiment reveals that constant $c_2 = -1.22$ is nearly identical for two different materials (Stainless Steel 304L and Aluminum 6061-T6) tested. Interestingly, $c_2 = -1.22$ holds for torsion, bending as well as rotating-bending. However, c_1 depends on the type of material and load (see Table 4.4). They showed that experimental data represented as the number of cycles to failure plotted against the initial slope of temperature rise can be consolidated into a single curve called a *universal curve*. Since the initial slope of the temperature evolution, R_θ, is evaluated at the very beginning of the number of cycles, the method provides a very rapid method for prediction of fatigue failure.

Figure 4.25 shows the universal curve for Steel 4330 and Steel 4145 subjected to torsion, bending, and rotating-bending fatigue. Results reveal that regardless of the stress state, the fatigue life against initial slope comparatively follows the universal curve for both steels and renders the slope of temperature rise a good candidate for fast prediction of fatigue failure.

A summary of the fatigue life predictive methodologies that employ temperature as an index of failure is given in Table 4.5 with a brief explanation of their applicability and challenges.

TABLE 4.4

Values for Coefficient c_1

Material	Torsion	Rotating–Bending	Bending
4330	7000	2350	1400
4145	9100	3500	3150

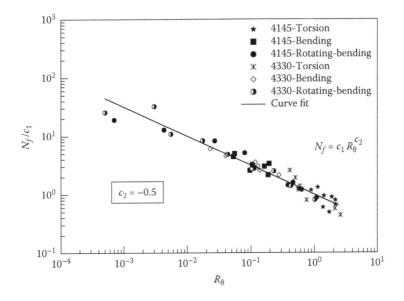

FIGURE 4.25 Fatigue life prediction based on initial temperature rise.

TABLE 4.5
Fatigue Life Predictive Methods Based on Temperature Evolution

Reference	Type of Loading	Failure Criteria	Applicability and Challenge(s)
Huang et al. (1984)	Rotating-bending	$R_\gamma = C' \cdot \exp\left(\dfrac{G}{(N_f)^{1/b}}\right)$	It provides a good indication of imminent final fracture. However, it is inapt for Phase 1 and Phase 2.
Jiang et al. (2001)	Uniaxial	$(N_f)^m = C \quad T$	It is applicable in Phase 2. However, it is inapt for some materials that do not exhibit steady-state temperature in Phase 2.
Fargione et al. (2002)	Uniaxial	$\phi = \displaystyle\int_0^{N_f} T\, dN = const.$	It is applicable to both LCF and HCF. However, it requires performing tests from the beginning to the final fracture.
Meneghetti (2007)	Uniaxial	$N_f = C\,(Q)^m$	It is applicable to both LCF and HCF. However, it is merely applicable to materials with constant energy dissipation per cycle.
Amiri and Khonsari (2010a, 2010b)	Torsion, bending	$N_f = c_1 R_\theta^{c_2}$	It provides a considerably fast prediction of failure.

In Chapter 6, we employ temperature evolution during the fatigue process to evaluate entropy generation and predict fatigue life based on accumulative entropy.

REFERENCES

Allen, D.H. 1985. A prediction of heat generation in a thermoviscoplastic uniaxial bar. *Int. J. Solids Struct.* 21, 325–342.

Amiri, M. and Khonsari, M.M. 2010a. Rapid determination of fatigue failure based on the temperature: Fully reversed bending load. *Int. J. Fatigue* 32, 382–389.

Amiri, M. and Khonsari, M.M. 2010b. Life prediction of metals undergoing fatigue load based on temperature evolution. *Mater. Sci. Eng. A* 527, 1555–1559.

Amiri, M., Naderi, M., and Khonsari, M.M. 2011. An experimental approach to evaluate the critical damage. *Int. J. Damage Mech.* 20, 89–112.

Ashby, M.F. and Jones, D.R.H. 2012. *Engineering Materials 1: An Introduction to Properties, Applications and Design*, 4th ed. Waltham, MA: Butterworth-Heinemann.

Backstrom, M. and Marquis, G. 2001. A review of multiaxial fatigue of weldments: Experimental results, design code and critical plane approaches. *Fatigue Frac. Eng. Mat. Struc.* 24, 279–291.

Bannantine, J.A., Comer, J.J., and Handrock, J.L. 1997. *Fundamentals of Metal Fatigue Analysis*. Englewood Cliffs, NJ: Prentice Hall.

Basquin, O.H. 1910. The exponential law of endurance tests. *Proc. ASTM, Part II*, 10, West Conshohocken, PA, 625–630.

Bathias, C. and Paris, P.C. 2004. *Gigacycle Fatigue in Mechanical Practice*. Boca Raton, FL: CRC Press, Taylor & Francis Group.

Blotny, R. and Kaleta, J. 1986. A method for determining the heat energy of fatigue process in metals under uniaxial stress. *Int. J. Fatigue* 8, 29–33.

Blotny, R., Kaleta, K., Grzebien, W., and Adamczewski, W. 1986. A method for determining the heat energy of fatigue process in metals under uniaxial stress. *Int. J. Fatigue* 8, 35–38.

Case, J., Chilver, A.H., and Ross, C.T.F. 1999. *Strength of Materials and Structures*, 4th ed. New York: John Wiley & Sons, Inc.

Charkaluk, E. and Constantinescu, A. 2009. Dissipative aspects in high cycle fatigue. *Mech. Materi.* 41, 483–494.

Dowling, N.E. 2004. Mean stress effects in stress-life and strain-life fatigue. SAE Fatigue 2004, *Proc. 3rd Int. SAE Fatigue Congress*, Sao Paulo, Brazil.

Eshelby, J.D. 1949a. Dislocations as a cause of mechanical damping in metals. *Proc. R. Soc. A* 197, 396–416.

Eshelby, J.D. 1949b. Uniformly moving dislocations. *Proc. Phys. Soc. A* 62, 307–314.

Fargione, G., Geraci, A., La Rosa, G., and Risitano, A., 2002. Rapid determination of the fatigue curve by the thermographic method. *Int. J. Fatigue* 24, 11–19.

Fatemi, A. and Yang, L. 1998. Cumulative fatigue damage and life prediction theories: A survey of the state of the art for homogeneous materials. *Int. J. Fatigue* 20, 9–34.

Feltner, C.E. and Morrow, J.D. 1961. Microplastic strain hysteresis energy as a criterion for fatigue fracture. *J. Basic Eng., Trans. ASME, Ser. D* 83, 15–22.

Fingers, R.W., Brown, W.F., Gross, B., and Srawley, J.E. 1980. Prediction model for fatigue crack growth in windmill structures. *ASTM Special Technical Publication*, ed. D.F., Bryan and J.M., Pottler, 714, 185–204.

Garud, Y.S. 1981. Multiaxial fatigue: A survey of the state-of-the-art. *J. Test Eval.* 9, 165–178.

Gerber, H. 1874. Bestimmung der zulassigen Spannungen in Eisenkonstructionen. *Z. Bayerischen Architeckten Ingenieur-Vereins*, 6, 101–110.

Ghoniem, N.M., Busso, E.P., Kioussis, N., and Huang, H. 2003. Multiscale modeling of nanomechanics and micromechanics: An overview. *Philosophical Magazine* 83, 3475–3528.

Glinka, G., Shen, G., and Plumtree, A. 1995. A multiaxial fatigue strain energy density parameter related to the critical plane. *Fatigue Fract. Eng. Mater. Struct.* 18, 37–46.

Golos, K.M. 1995. A total strain energy density model of metal fatigue. *Strength Mater.* 27, 32–41.

Goodman, J. 1919. *Mechanics Applied to Engineering.* London: Longmans, Green and Co., 631–636.

Gross, T. 1981. Calorimetric of measurement of the plastic work of fatigue crack propagation in quenched and tempered 4140 steel. Ph.D. thesis, Northwestern University.

Halford, G.R. 1966. The energy required for fatigue. *ASTM J. Mater.* 1, 3–18.

Hirth, J.P. and Lothe, J. 1968. *Theory of Dislocations.* New York: McGraw-Hill Inc.

Hodowany, J. 1997. On the conversion of plastic work into heat. Ph.D. thesis, California Institute of Technology.

Huang, Y., Li, S.X., Lin, S.E., and Shih, C.H. 1984. Using the method of infrared sensing for monitoring fatigue process of metals. *Mater. Eval.* 42, 1020–1024.

Jiang, L., Wang, H., Liaw, P.K., Brooks, C.R., and Klarstrom, D.L. 2001. Characterization of the temperature evolution during high-cycle fatigue of the ULTIMET superalloy: Experiment and theoretical modeling. *Metall. Mater. Trans. A* 32, 2279–2296.

Kaechele, L. 1963. *Review and Analysis of Cumulative-Fatigue-Damage Theories.* RM-3650-PR, Santa Monica, CA: The Rand Corporation.

Kaleta, J., Blotny, R., and Harig, H. 1990. Energy stored in a specimen under fatigue limit loading conditions. *J. Testing Eval. JETVA* 19, 326–333.

Kapoor, R. and Nemat-Nasser, S. 1998. Determination of temperature rise during high strain rate deformation. *Mech. Mater.* 27, 1–12.

Karolczuk, A. and Macha, E. 2005. A review of critical plane orientations in multiaxial fatigue failure criteria of metallic materials. *Int. J. Fract.* 134, 267–304.

Kuhlman-Wilsdorf, D. 1987. LEDS: Properties and effects of low energies dislocation structures. *Mater. Sci. Eng.* 86, 53–66.

Kujawski, D. and Ellyin, F. 1995. A unified approach to mean stress effect on fatigue threshold conditions. *Int. J. Fatigue* 17, 101–106.

Langer, B.F. 1937. Fatigue failure from stress cycles of varying amplitude. *Trans. Am. Soc. Mech. Engrs.* 59, A-160.

Lawson, A.W. 1941. The effect of stress on internal friction in polycrystalline copper. *Physc. Rev.* 60, 330–335.

Lukas, P. and Kunz, L. 1999. Specific features of high-cycle and ultra-high-cycle fatigue. *Fatigue Fract. Eng. Mater. Struct.* 22, 747–753.

Luong, M.P. 1998. Fatigue limit evaluation of metals using an infrared thermographic technique. *Mech. Mater.* 28, 155–163.

Manson, S.S. and Halford, G.R. 2005. *Fatigue and Durability of Structural Materials.* Materials Park, OH: ASM International.

McDowell, D.L. 2008. Viscoplasticity of heterogeneous metallic materials. *Mater. Sci. Eng. R: Reports* 62, 67–123.

Meneghetti, G. 2006. Analysis of the fatigue strength of a stainless steel based on the energy dissipation. *Int. J. Fatigue* 29, 81–94.

Meneghetti, G. 2007. Analysis of the fatigue strength of a stainless steel based on the energy dissipation. *Int. J. Fatigue* 29, 81–94.

Min, L., Xiaofei, X., and Qing-Xiong, Y. 1995. Cumulative fatigue damage dynamic interference statistical model. *Int. J. Fatigue* 17, 559–566.

Miner, M.A. 1945. Cumulative damage in fatigue. *J. Appl. Mech.* 14, 159–164.

Moore, J.T. and Kuhlman-Wilsdorf, D. 1970. The rate of energy storage in workhardened metals. *2nd Int. Conf. Strength Metals Alloys*, 2, Metals Park, OH: American Society of Metals, 484–488.

Morrow, J.D. 1965. Cyclic plastic strain energy and fatigue of metals. In *Internal Friction, Damping, and Cyclic Plasticity, ASTM STP*, Philadelphia, PA, 378, 45–84.

Mughrabi, H. 1999. On the life-controlling microstructural fatigue mechanisms in ductile metals and alloys in the gigacycle regime. *Fatigue Fract. Eng. Mater. Struct.* 22, 633–641.

Mughrabi, H. 2009. Cyclic slip irreversibilities and the evolution of fatigue damage. *Metall. Mater. Trans. A* 40, 1257–1279.

Mura, T. and Lyons, W.C. 1966. Continuous distribution of dislocations and energy dissipation in metals. *J. Acoust. Soc. Am.* 39, 527–531.

Nabarro, F.R.N. 1967. *Theory of Crystal Dislocations.* Oxford, UK: Oxford University Press.

Newmark, N.M. 1952. A review of cumulative damage in fatigue. In *Fract, Fatigue Metals,* ed. W.M. Murray, 197–228. New York: The Technology Press of the MIT-Wiley.

Nishijima, S. and Kanazawa, K. 1999. Stepwise S–N curve and fish-eye failure in gigacycle fatigue. *Fatigue Fract. Eng. Mater. Struct.* 22, 601–607.

Nosonovsky, M. and Bhushan, B. 2007. Multiscale friction mechanisms and hierarchical surfaces in nano- and bio-tribology. *Mater. Sci. Eng. R: Reports* 58, 162–193.

Ozisik, M.N. 2002. *Boundary Value Problems of Heat Conduction.* Mineola, NY: Dover Publications.

Padilla II, H.A., Smith, C.D., Lambros, J., Beaudoin, A.J., and Robertson, I.M. 2007. Effects of deformation twinning on energy dissipation in high rate deformed zirconium. *Metall. Mater. Trans. A* 38, 2916–2927.

Palmgren, A.. 1924. The fatigue life of ball bearing [in German]. *Zeitschrift des Vereines Deutscher Ingenieure,* 68, 339–341.

Papadopoulos, I.V., Avoli, P., Gorla, C., Filippini, M., and Bernasconi, A. 1997. A comparative study of multiaxial high-cycle fatigue criteria for metals. *Int. J. Fatigue* 19, 219–235.

Park, J. and Nelson, D. 2000. Evaluation of an energy-based approach and a critical plane approach for predicting constant amplitude multiaxial fatigue life. *Int. J. Fatigue* 22, 23–39.

Pook, L.P. 2007. *Metal Fatigue: Why It Is, Why It Matters.* Dordrecht, The Netherlands: Springer.

Ranc, N., Wagner, D., and Paris, P.C. 2008. Study of thermal effects associated with crack propagation during very high cycle fatigue tests. *Acta Mater.* 56, 4012–4021.

Schijve, J. 2001. *Fatigue of Structures and Materials.* Dordrecht, The Netherlands: Kluwer Academic Publishers.

Schwarts, R.T. 1948. Correlation of data on the effect of range of stress on the fatigue strength of metals for tensile mean stresses. Master's thesis, Ohio State University.

Shigley, J.E. and Mitchell, L.D. 1983. *Mechanical Engineering Design,* 4th ed. New York: McGraw Hill.

Skelton, R.P. 2004. Hysteresis, yield, and energy dissipation during thermo-mechanical fatigue of a ferritic steel. *Int. J. Fatigue* 26, 253–264.

Smith, K.N., Watson, P., and Topper, T.H. 1970. A stress–strain function for the fatigue of metals. *J. Mater.* 5, 767–78.

Socie, D. 1993. Critical plane approaches for multiaxial fatigue damage assessment, In *Advances in Multiaxial Fatigue,* ASTM STP 1191, ed. D.L. McDowell and R. Ellis. Philadelphia, PA: Am. Soc. Testing Mater., 7–36.

Socie, D.F., and Marquis, G.B. 2000. *Multiaxial Fatigue.* Warrendale, PA: Soc. Autom. Engrs.

Soderberg, C.R. 1939. Factor of safety and working stress. *Trans. Am. Soc. Mech. Eng.* 52, 13–28.

Stanfield, G. 1935. Discussion on the strength of metals under combined alternating stresses, ed. H. Gough and H. Pollard, *Proc. Inst. Mech. Engrs.* 131, 3–103.

Stephens, R.I., Fatemi, A., Stephens, R.R., and Fuchs, H.O. 2000. *Metal Fatigue in Engineering,* 2nd ed. New York: John Wiley & Sons, Inc.

Stroh, A.N. 1953. A theoretical calculation of the stored energy in a work-hardened material. *Proc. R. Soc. A* 218, 391–400.

Subhash, G., Ravichandran, G., and Pletka, B.J. 1997. Plastic deformation of hafnium under uniaxial compression. *Metall. Mater. Trans. A* 28, 1479–87.

Szczepanski, C.J., Jha, S.K., Larsen, J.M., and Jones, J.W. 2008. Microstructural influences on very-high-cycle fatigue-crack initiation in Ti-6246. *Metall. Mater. Trans. A* 39, 2841–2851.

Tabanli, R.M., Simha, N.K., and Berg, B.T. 1999. Mean stress effects on fatigue of NiTi. *Mater. Sci. Eng. A* 273–275, 644–648.

Walker, K. 1970. The effect of stress ratio during crack propagation and fatigue for 2024-T3 and 7075-T6 aluminum, *Effects of Environment and Complex Load History on Fatigue Life*, ASTM STP 462. Philadelphia, PA: Am. Soc. for Testing Mater., 1–14.

Wang, H., Jiang, L., Brooks, C.R., and Liaw, P.K.. 2000. Infrared temperature mapping of ULTIMET alloy during high-cycle fatigue tests. *Metall. Mater. Trans.* A 31, 1307–1310.

Wang, Y.Y. and Yao, W.X. 2004. Evaluation and comparison of several multiaxial fatigue criteria. *Int. J. Fatigue* 26, 17–25.

Woodward, A.R., Gunn, K.W., and Forrest, G. 1956. The effect of mean stress on the fatigue of aluminum alloys. *Int. Conf. Fatigue Metals*, London, Institution of Mechanical Engineers, Westminster, London, 1156–1158.

You, B.R. and Lee, S.B. 1996. A critical review on multiaxial fatigue assessments of metals. *Int. J. Fatigue* 18, 235–44.

Zener, C. 1937. Internal friction in solids I: Theory of internal friction in reeds. *Phys. Rev.* 52, 230–235.

Zener, C. 1938. Internal friction in solids II: General theory of thermoelastic internal friction. *Phys. Rev.* 53, 90–99.

Zener, C. 1940. Internal friction in solids. *Proc. Phys. Soc.* 52, 152–166.

Zener, C. 1948. *Elasticity and Anelasticity of Metals*. Chicago: The University of Chicago Press.

Zhang, G.P. and Wang, Z.G. 2008. Fatigue of small-scale metal materials: From micro to nano-scale. In *Multiscale Fatigue Crack Initiation and Propagation of Engineering Materials: Structural Integrity and Microstructural Worthiness*, ed. G.C. Sih. New York: Springer Science + Business Media, B.V.

5 Basic Thermodynamic Framework for Fatigue Analysis

In this chapter, we study the deformation characteristics of a solid material subjected to fatigue loading. We are particularly interested in determining a material's degradation characteristics in response to entropy generation and accumulation resulting in degradation. In Chapter 4, we defined stress and strain in a mechanical framework without bringing into play any thermodynamics considerations. In this chapter, we proceed to formulate the response of a material to the applied stresses within the context of thermodynamics and show that it provides a useful framework for studying mechanical fatigue.

The study of elastic-plastic deformation of solids within the thermodynamic framework dates back to over half a century ago. Biot (1955, 1956, 1957, 1958, 1973) developed original concepts and put forward the theory of thermoelasticity within an irreversible thermodynamic context. Assuming that an elastic solid is initially at equilibrium state, he introduced the concept of thermoelastic dissipation function in terms of the entropy flow. The dissipation function characterizes the internal damping in elastic bodies and represents the irreversibility in the medium. Irreversibility—as a measure of departure from the ideal reversible process (Keenan 1951)—is always accompanied by entropy generation and increase in disorder.

Let us consider a system with geometry as shown in Figure 5.1. The total volume of the system is V and the boundary of the system, Ω, is designated by the dashed lines. The system is in part exposed to the surroundings, and the fixed boundary shown in Figure 5.1 designates the location to which the system is anchored. This could be the gripped section of a laboratory specimen or represent a part of the system in contact with another component that restrains its motion. Time-dependent, cyclic fatigue load, $F(t)$, is externally applied to the system. To analyze the thermodynamical behavior of this system, the following assumptions are made.

(1) Heat exchange with the surroundings takes place through convection and radiation mechanisms and there is no mass transfer;
(2) The only heat transfer mechanism through the fixed boundary is by conduction;
(3) The variation of kinetic energy, KE, and potential energy, PE, is negligible;
(4) The work done by external forces, such as gravity, is negligible;
(5) Diffusion of matter within the deformed system is negligible;
(6) The density of the system remains constant; and
(7) The concept of local equilibrium holds.

We begin by writing the statement of energy conservation. In accordance with the thermodynamic formulations developed in Chapter 2, the total energy content, E, within an

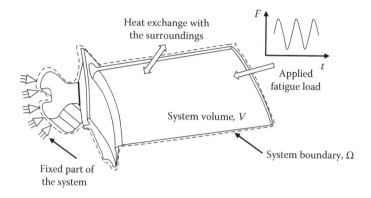

FIGURE 5.1 A schematic of a fatigue system with imposed boundary and external conditions.

arbitrary volume, ∂V, of a differential element changes if energy flows into (or out of) the volume through its boundary $\partial \Omega$. That is,

$$\frac{dE}{dt} = \dot{E}_{in} - \dot{E}_{out} \tag{5.1}$$

where dE/dt is the rate of total energy and $\dot{E}_{in} - \dot{E}_{out}$ is the rate of net energy exchange by heat and work. The dot denotes differentiation with respect to time, t. Based on assumptions (3) and (4) and considering Equation (2.2), the change in the total energy, E, and the internal energy, U, are identical. That is

$$\frac{dE}{dt} = \frac{dU}{dt} \tag{5.2}$$

Equation (5.2) can be written in terms of specific quantities as

$$\frac{d(\rho e)}{dt} = \frac{d(\rho u)}{dt} \tag{5.3}$$

where e is the total specific energy, u is the specific internal energy, and ρ is the density. Substituting Equation (5.3) into Equation (5.1) and considering the integral form of Equation (5.1), we arrive at the following form of the conservation of energy:

$$\frac{d}{dt}\int_V \rho u\, dV = \rho \int_V \frac{\partial u}{\partial t}\, dV = -\int_\Omega J_e \cdot d\Omega \tag{5.4}$$

where J_e is the total energy flux. Employing Gauss's theorem, the local form of the law of conservation of energy can be obtained. The result is (De Groot and Mazur 1962)

$$\rho \frac{\partial u}{\partial t} = - \nabla \cdot J_e \tag{5.5}$$

As discussed in Chapter 2, one can start with Equation (5.5) and derive the balance equation for the specific internal energy (the first law of thermodynamics) as follows:

$$\rho\frac{du}{dt} = \sigma:D - \nabla \cdot J_q \tag{5.6}$$

where J_q is the heat flux, σ is the symmetric stress tensor, and D is the symmetric rate of the deformation tensor. In what follows, we show how the first law can be expressed in terms of measurable variables.

Example 5.1

Consider the Fourier's law of heat diffusion and assume that the system under-goes small deformation. Rewrite the first law of thermodynamics, Equation (5.6), in terms of measurable variables, that is, strain ε and temperature T.

SOLUTION

Based on Fourier's law of heat conduction, the heat flux vector J_q is linearly related to the temperature gradient vector ∇T:

$$J_q = -k\nabla T \tag{5.7}$$

where k is the thermal conductivity of the material. If the thermal conductivity is uniform throughout the body, then, taking divergence of Equation (5.7) yields

$$\nabla \cdot J_q = -k\nabla^2 T \tag{5.8}$$

The rate of change of internal energy for a solid material can be replaced by

$$\rho\frac{du}{dt} = \rho c\frac{dT}{dt} \tag{5.9}$$

in which c is either the specific heat at constant pressure or constant volume. It is known (Lemaitre and Chaboche 1990) that for small deformations, the rate of change of deformation tensor D can be replaced by the rate of strain $\dot{\varepsilon}$. Therefore, Equation (5.6) can be written as

$$\rho c\frac{dT}{dt} = \sigma:\dot{\varepsilon} + k\nabla^2 T \tag{5.10}$$

where, of course, $\sigma:\dot{\varepsilon} = \sigma_{ij}\dot{\varepsilon}_{ij}$. Equation (5.10) is the statement of the energy balance equation for a solid material undergoing deformation in terms of mea-surable quantities T, σ, ε, and time t with the assumption of constant ρ, c, and k. ▲

The total strain tensor can be decomposed into elastic and plastic strains as

$$\varepsilon_{ij} = \varepsilon_{ij}^{e} + \varepsilon_{ij}^{p} \tag{5.11}$$

The elastic part of the above decomposition is the strain generated by the applied stress if the deformation is merely elastic; that is, once the stress vanishes, the solid returns to its initial state and the process is reversible. Conversely, the plastic deformation is accompanied by irreversibility resulting in entropy production. We shall first examine some useful aspects of elastic strain within the thermodynamic framework before proceeding to study the role of plastic deformation in entropy production.

5.1 ENTROPY BALANCE EQUATION OF A DEFORMED BODY

Figure 5.2 shows a system that experiences deformation via an external element such as mechanical stress, thermal stress, and electrical field. The total entropy is a combination of entropy flow across the boundary of the body and entropy production within the system.

$$dS = d_{e}S + d_{i}S \tag{5.12}$$

At very low stress amplitudes, the material behavior is elastic and the contribution of entropy generation $d_{i}S$ in Equation (5.12) is nil. On the other hand, at higher stress levels, plastic deformation becomes significant resulting in permanent (irreversible) changes to the microstructure of the materials. In fact, the results of low-cycle fatigue reveal that in this case, entropy generation term, $d_{i}S$, is dominated by the plastic deformation (Naderi, Amiri, and Khonsari 2010). The alteration of microstructure during plastic deformation mainly corresponds to the creation of the dislocations. Thus, there is no irreversibility if deformation is purely elastic and dislocations are locked (Beghi 1982).

It is common in thermoelasticity to deal with free energy ψ rather than internal energy u. The Helmholtz free energy is defined as

$$\psi = u - Ts \tag{5.13}$$

As we learned in Chapter 3, the Helmholtz free energy—also referred to as the available energy—represents the amount of energy available for doing useful work (Crowe and Feinberg 2001). The free energy ψ decreases with time as the fatigue process continues and the material degrades. Therefore, ψ is a concave function with respect to temperature T

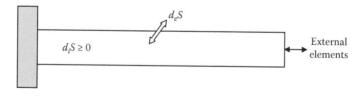

FIGURE 5.2   A system subjected to external elements.

and convex with respect to other variables. Through the following example, we show how the Helmholtz free energy can be used to simplify the general form of equations derived in the preceding section.

Example 5.2

Starting from the conservation of energy, Equation (5.6), derive the appropriate equations for relating the stress-strain law and entropy to Helmholtz free energy. Assume that the free energy function depends only on temperature T and elastic strain ε_e.

Solution

Rewriting Equation (5.6) and making use of Equation (5.11) yields

$$\rho \frac{du}{dt} = \sigma : \dot{\varepsilon}_p + \sigma : \dot{\varepsilon}_e - \quad \cdot J_q \tag{5.14}$$

Taking the time derivative of Equation (5.13), we have

$$\frac{d\psi}{dt} = \frac{du}{dt} - s \frac{dT}{dt} - T \frac{ds}{dt} \tag{5.15}$$

Upon elimination of du/dt between Equation (5.15) and Equation (5.14), we have

$$\rho \left(\frac{d\psi}{dt} + s\dot{T} + T\dot{s} \right) = \sigma : \dot{\varepsilon}_p + \sigma : \dot{\varepsilon}_e - \quad \cdot J_q \tag{5.16}$$

Since the free energy function is assumed to depend on temperature and elastic strain, we have

$$\psi = \psi(T, \varepsilon_e) \tag{5.17}$$

Taking the time derivative of Equation (5.17) yields

$$\frac{d\psi}{dt} = \frac{\partial \psi}{\partial \varepsilon_e} : \dot{\varepsilon}_e + \frac{\partial \psi}{\partial T} \dot{T} \tag{5.18}$$

Substituting Equation (5.17) into Equation (5.15) and rearranging the resultant equation, we have

$$\left(\rho \frac{\partial \psi}{\partial \varepsilon_e} - \sigma \right) : \dot{\varepsilon}_e + \rho \left(\frac{\partial \psi}{\partial T} + s \right) \dot{T} - \sigma : \dot{\varepsilon}_p + \quad \cdot J_q + \rho T \dot{s} = 0 \tag{5.19}$$

Since Equation (5.19) is valid for any arbitrary $\dot{\varepsilon}_e$ and $\dot{T}$, the following equation should hold.

$$\rho \frac{\partial \psi}{\partial \varepsilon_e} = \sigma \tag{5.20a}$$

$$-\frac{\partial \psi}{\partial T} = s \tag{5.20b}$$

$$-\sigma : \dot{\varepsilon}_p + \quad \cdot J_q + \rho T \dot{s} = 0 \tag{5.20c}$$

Equation (5.20a) describes the relationship between stress and strain tensors, and Equation (5.20b) describes how entropy can be obtained by differentiating the Helmholtz function with respect to temperature. Equations (5.20a) and (5.20b) are known as the *constitutive equations of materials* (Parkus 1968). Equation (5.20c) defines the balance of entropy of the system.▲

Equation (5.20c) can be developed further to arrive at useful conclusions. Starting from Equation (5.20b) and making use of Equation (5.18) one can write the rate of entropy as

$$\dot{s} = -\frac{d}{dt}\left(\frac{\partial \psi}{\partial T}\right) = -\frac{\partial^2 \psi}{\partial \varepsilon^e \partial T} : \dot{\varepsilon}_e - \frac{\partial^2 \psi}{\partial T^2}\dot{T} \tag{5.21}$$

Substitution of Equations (5.20a) and (5.20b) into Equation (5.21) yields

$$\dot{s} = -\frac{1}{\rho}\frac{\partial \sigma}{\partial T} : \varepsilon_e + \frac{\partial s}{\partial T}\dot{T} \tag{5.22}$$

Now, $\partial s/\partial T$ can be replaced by c/T where c is the specific heat capacity. Substituting Equation (5.22) into Equation (5.20c), we arrive at

$$\rho c \dot{T} + \quad \cdot J_q - \sigma : \dot{\varepsilon}_p - T\frac{\partial \sigma}{\partial T} : \dot{\varepsilon}_e = 0 \tag{5.23}$$

Using Equation (5.8), Equation (5.23) can be written in the following form:

$$\rho c \dot{T} = k \quad ^2 T + \sigma : \dot{\varepsilon}_p + T\frac{\partial \sigma}{\partial T} : \dot{\varepsilon}_e \tag{5.24}$$

Equation (5.24) is similar to the energy balance Equation (5.10) with the difference that in Equation (5.24) the energy associated with plastic and elastic deformation is explicitly expressed. Equation (5.24) shows the balance of rate of energy between four terms: the rate of change of internal energy $\left(\rho c \dot{T}\right)$, which is mainly due to conduction

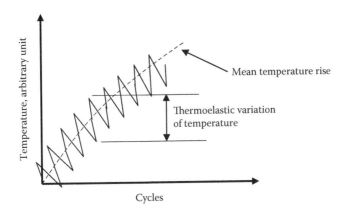

FIGURE 5.3   Mean temperature rise and temperature variation due to thermoelastic effect.

of heat inside the solid material ($k\,^2T$); plastic energy dissipation ($\sigma : \dot{\varepsilon}_p$), responsible for mean temperature rise; and the thermoelastic coupling term ($T\,\partial\sigma/\partial T : \dot{\varepsilon}_e$), responsible for fluctuation of temperature around the mean value. Temperature variation due to plastic energy dissipation and thermoelastic effect is shown schematically in Figure 5.3.

Equation (5.24) is of great importance in the analysis of elastic bodies undergoing thermal stress. To illustrate, let us borrow a page from the thermoelastic theory for an isotropic solid (Nowaki 1962). The stress and strain equations for an isotropic thermoelastic solid are given as

$$\sigma_{ij} = \frac{E}{1+\nu}\left(\varepsilon_{ij} + \frac{\nu}{1-2\nu}\varepsilon_{kk}\delta_{ij}\right) - \frac{E}{1-2\nu}\alpha(T-T_0)\delta_{ij} \quad \text{(Hooke's Law)} \qquad (5.25)$$

$$\varepsilon_{ij} = \frac{1+\nu}{E}\sigma_{ij} - \frac{\nu}{E}\sigma_{kk}\delta_{ij} + \alpha(T-T_0)\delta_{ij} \qquad (5.26)$$

where σ_{ij} is the stress tensor, ε_{ij} is the strain tensor, E represents the elastic modulus, ν denotes the Poisson's ratio, α is the linear thermal expansion coefficient, δ_{ij} is the Kronecker delta, and T_0 is the reference temperature. Substituting Equation (5.25) into Equation (5.24) and assuming that there is no plastic deformation, $(\sigma : \dot{\varepsilon}_p = 0)$, we arrive at the following equation:

$$\rho c \dot{T} = k\,^2T - \frac{E\alpha}{1-2\nu}T\dot{\varepsilon}_{kk} \qquad (5.27)$$

Equation (5.27) is known as the *coupled heat conduction equation* involving the coupling term, $E\alpha T\dot{\varepsilon}_{kk}/(1-2\nu)$, which includes both temperature and strain. In practical engineering applications, it is common to first neglect the thermoelastic coupling term in Equation (5.27) and solve the regular heat conduction equation to find the temperature field regardless of stress. Stress and strains are then evaluated once the temperature field is known.

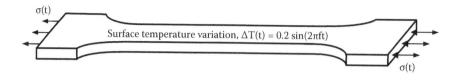

FIGURE 5.4   Uniaxial stress and temperature variation of a sample.

Example 5.3

Suppose that a sinusoidally varying uniaxial load, $\sigma(t)$, is applied to a specimen as shown in Figure 5.4. The specimen is made of Aluminum 6061-T6 with the density $\rho = 2700$ (kg/m³), thermal expansion coefficient $\alpha = 23.4$ (10^{-6}/K), and the specific heat capacity $c = 896$ (J/kg.K). The reference temperature is taken at room temperature, $T_0 = 25°C$. Assume that the material deforms within the elastic range and that small variation of temperature, $\Delta T = T-T_0$, due to thermoelastic effect can be approximated by $\Delta T = 0.2 \sin(2\pi ft)$, where f denotes the frequency. This temperature variation function is obtained by applying a curve-fitting routine to the temperature output of an infrared (IR) camera (Giancane et al. 2009). Determine the change in stress, $\sigma(t)$, associated with this temperature variation. Treat the problem as one-dimensional and assume surface temperature is uniform.

SOLUTION

We start by employing Equation (5.27). Since the temperature is assumed to be uniform, $\nabla^2 T = 0$. Within the elastic range, the variation in temperature is very small, so that temperature T can be replaced by reference temperature T_0. Therefore, for small change in temperature ΔT, Equation (5.27) can be written as

$$\dot{T} = -\frac{ET_0\alpha}{\rho c(1-2v)}\ \dot{\varepsilon}_{kk} \tag{5.28}$$

The term involving $\Delta\varepsilon_{kk}$ can be evaluated from Equation (5.26) for small variation in temperature as

$$\varepsilon_{kk} = \frac{1+v}{E}\ \sigma_{kk} - \frac{3v}{E}\ \sigma_{kk} = \frac{1-2v}{E}\ \sigma_{kk} \tag{5.29}$$

Substitution of Equation (5.29) into Equation (5.28) yields

$$\dot{T} = -\frac{\alpha T_0}{\rho c}\ \dot{\sigma}_{kk} \tag{5.30}$$

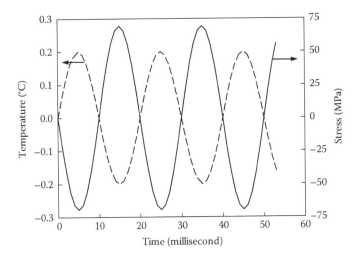

FIGURE 5.5 Stress measurement based on temperature readings

Since the problem is one-dimensional, the stress variation can be evaluated from Equation (5.30) as

$$\sigma = -\frac{\rho c}{\alpha T_0} \; T = -\frac{2700 \times 896}{23.4 \times 10^{-6} \times 298.15} \times (0.2)\, \sin(2\pi f t)$$

$$= -69.35 \sin(2\pi f t)\,\text{MPa} \qquad\qquad (5.31)$$

Figure 5.5 depicts the variation of stress evaluated from temperature measurement. The variation in temperature is also plotted in this figure. Equation (5.30) provides the correlation that researchers rely on to evaluate the stress field from temperature measurements. Thanks to IR thermography technique, it is possible to capture a very small variation in temperature due to elastic deformation. Equation (5.30) shows that when a specimen is exposed to tension, its temperature drops, and it heats when subjected to compression. This is the meaning of the negative sign in Equation (5.30). ▲

5.2   ENTROPY CHANGE DUE TO THERMAL DEFORMATION

As the system exchanges thermal energy with the surroundings, its entropy can be affected due to the change in the temperature field as well as the thermally induced strain since both the temperature gradient within the solid body and the associated thermal strain can contribute to the change of entropy. To illustrate, consider a body subjected to an external thermal force. The external thermal force can be introduced into the system, for example, by changing the environment temperature as shown in Figure 5.6.

Assuming that the volume change is small enough such that a material elongates in the elastic range, we wish to investigate the change in entropy of the system, $s = s(v, T)$, with v denoting specific volume, $v = 1/\rho$. The change in the entropy is simply:

$$ds = \left(\frac{\partial s}{\partial v}\right)_T dv + \left(\frac{\partial s}{\partial T}\right)_v dT \qquad\qquad (5.32)$$

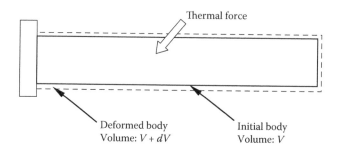

FIGURE 5.6 Deformation of solid material due to thermal stress.

The following correlations for a solid material are well known (Sonntag, Borgnakke, and Van Wylen 2003):

$$\left(\frac{\partial s}{\partial T}\right)_v = \frac{c_v}{T} \tag{5.33}$$

$$\left(\frac{\partial s}{\partial v}\right)_T = -\frac{\alpha}{\kappa_T} \tag{5.34}$$

where c_v is the specific heat capacity at constant volume and α and κ_T are the coefficients of thermal expansion and compressibility, respectively. Substitution of Equation (5.33) and Equation (5.34) into Equation (5.32) results in the following equation for the variation of entropy of the solid material due to thermal expansion and temperature variation:

$$ds = -\frac{\alpha}{\kappa_T} dv + c_v \frac{dT}{T} \tag{5.35}$$

Equation (5.35) can be integrated from an initial condition (say, a strain-free solid at room temperature) up to the desired temperature rise. The first term on the right-hand side of the above equation is associated with the thermal strain (expansion) and the second term is due to the temperature variation, which can be obtained by solving the heat conduction equation. A one-dimensional numerical example of the entropy measurement is given in the work of Al Nassar (2003) for a steel material. It is worthwhile to note that the solution to Equation (5.35) gives the change in total entropy of the system, not necessarily the entropy generation.

Example 5.4

An isotropic sphere of mass m and specific heat capacity c initially at temperature $T = T_i$ is immersed into a constant temperature bath at temperature $T_b > T_i$, see Figure 5.7. Employ the lumped capacitance analysis to arrive at an expression for the transient change of entropy associated with the thermal expansion of the sphere. Assume that the convection heat transfer coefficient from the sphere surface to the liquid, h, is known.

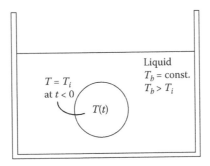

FIGURE 5.7   Solid sphere in a constant temperature bath.

SOLUTION

According to the lumped capacitance model, the temperature of the sphere is spatially uniform during the transient process. The energy balance equation between the sphere and the surrounding fluid is

$$mc\frac{dT}{dt} = -hA(T - T_b) \tag{5.36}$$

where $A(t)$ is the surface area of the sphere. Introducing temperature differences as

$$\theta = T_b - T \tag{5.37}$$

$$\theta_i = T_b - T_i \tag{5.38}$$

it follows that

$$\frac{mc}{Ah}\frac{d\theta}{\theta} = -dt \tag{5.39}$$

For an isotropic material, the increase in surface area due to thermal expansion is given by

$$A = A_i[1 + 2\alpha(T - T_i)] = A_i(B - 2\alpha\,\theta) \tag{5.40}$$

where A_i is the surface area at initial temperature T_i and $B = 1 + 2\alpha\,\theta_i$. Substituting Equation (5.40) into Equation (5.39) yields

$$\frac{mc}{A_i h}\frac{d\theta}{(B - 2\alpha\,\theta)\theta} = -dt \tag{5.41}$$

Integrating Equation (5.41) from the initial condition ($t = 0$, $\theta = \theta_i$) to an arbitrary time t yields

$$\frac{mc}{A_i h}\int_{\theta_i}^{\theta}\frac{d\theta}{(B - 2\alpha\,\theta)\theta} = -\int_{0}^{t} dt \tag{5.42}$$

Evaluating the integrals yields

$$\frac{\theta}{\theta_i} = \frac{T - T_b}{T_i - T_b} = \frac{e^{-\frac{BhA_i}{mc}t}}{1 + 2\alpha\, e^{-\frac{BhA_i}{mc}t}} \tag{5.43}$$

Equation (5.43) provides a relation for the temperature change with time as a function of physical parameters including thermal expansion. Similar to Equation (5.40), the change in volume of an isotropic material is given as

$$V = V_i[1 + 3\alpha\,(T - T_i)] \tag{5.44}$$

where V_i is the volume at initial temperature T_i. It follows from Equation (5.44) that

$$dv = \frac{dV}{m} = \frac{3\alpha\, V_i}{m}dT \tag{5.45}$$

where v is the specific volume.

Substituting Equation (5.45) into Equation (5.35) yields

$$ds = \left(-\frac{3\alpha^2\, V_i}{m\kappa_T} + \frac{c_v}{T}\right)dT \tag{5.46}$$

Integrating Equation (5.46) from the initial condition ($t = 0$, $s = s_i$) to an arbitrary temperature, we obtain

$$s = s - s_i = \int_{s_i}^{s} ds = \int_{T_i}^{T}\left(-\frac{3\alpha^2\, V_i}{m\kappa_T} + \frac{c_v}{T}\right)dT \tag{5.47}$$

Evaluating the integrals yields

$$s = \frac{3\alpha^2\, V_i}{m\kappa_T}(\theta - \theta_i) + c_v \ln\left(\frac{T_b - \theta}{T_b - \theta_i}\right) \tag{5.48}$$

Equation (5.48) gives the change in entropy of the sphere as it expands in the liquid. To show how the entropy of the sphere changes during the thermal expansion, we denote the first term on the right-hand side of Equation (5.48) by Δs_v, which shows the entropy change caused by volume expansion. Similarly, we represent the second term by Δs_T, which indicates the entropy change due to temperature variation. Therefore, Equation (5.48) can be written as

$$s = s_v + s_T \tag{5.49}$$

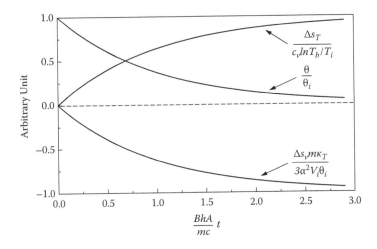

FIGURE 5.8 Evolution of change in entropy and temperature of the sphere for $T_i = 25°C$ and $T_b = 40°C$.

where

$$s_v = c_v \ln\left(\frac{T_b - \theta}{T_b - \theta_i}\right)$$

$$s_T = \frac{3\alpha^2 V_i}{m\kappa_T}(\theta - \theta_i) \tag{5.50}$$

Figure 5.8 shows the evolution of each of the contributions of entropy change (i.e., Δs_v and Δs_T) along with the evolution of the temperature. Note that for comparison purposes, the temperature and entropy changes are normalized as shown in the figure. It can be seen from Figure 5.8 that the normalized change in entropy due to temperature Δs_T increases by time, whereas, the normalized entropy change due to volume expansion Δs_v decreases. ▲

Entropy analysis of a thermally deformed body as illustrated in this example is a useful tool for the analysis of irreversibility associated with the change in temperature and density (volume). The spherical geometry considered in Example 5.4 is of interest in many applications such as thermodynamic analysis of thermocouple junction in a stream of fluid or heat treating of balls used in rolling element bearings. A similar analysis can be performed for cooling and for different geometries.

5.3 CLAUSIUS–DUHEM INEQUALITY

The entropy balance equation under the hypothesis of local equilibrium is given by Equation (2.32) and repeated below

$$\rho\frac{ds}{dt} = \dot{\gamma} - \nabla \cdot \boldsymbol{J}_{s,tot} \tag{5.51}$$

where $\dot{\gamma}$ denotes the entropy production per unit volume per unit time[*] and $\boldsymbol{J}_{s,tot}$ is the total *entropy flow* per unit area and per unit time. The second law of thermodynamics asserts that

$$\dot{\gamma} \geq 0 \qquad (5.52)$$

For a reversible process $\dot{\gamma} = 0$, while for an irreversible process the entropy production is non-negative. In processes involving low-cycle fatigue, damage occurs primarily as a result of local accumulation of plastic strain energy. Since plastic deformation is irreversible, it must be accompanied by irreversible production of entropy, that is, $\dot{\gamma} > 0$.

Assuming that the gradient of chemical potential induced by deformation and diffusion of matter within the body is negligible, the total entropy flow per unit area and per unit time simply reduces to the heat flux, viz.,

$$J_{s,tot} = \frac{J_q}{T} \qquad (5.53)$$

Upon the substitution of Equation (5.53) into Equation (5.51), we obtain

$$\rho \frac{ds}{dt} = \dot{\gamma} - \quad \cdot \left(\frac{J_q}{T} \right) \qquad (5.54)$$

The second term on the right-hand side of Equation (5.54) can be written as

$$\cdot \left(\frac{J_q}{T} \right) = \frac{\cdot J_q}{T} - J_q \cdot \frac{T}{T^2} \qquad (5.55)$$

Substituting Equation (5.55) into Equation (5.54) yields

$$\rho T \dot{s} = T \dot{\gamma} - \quad \cdot J_q + J_q \cdot \frac{T}{T} \qquad (5.56)$$

Making use of Equation (5.56) in Equation (5.20c) and rearranging the resulting equation for $\dot{\gamma}$ gives

$$\dot{\gamma} = \frac{\sigma : \dot{\varepsilon}_p}{T} - J_q \cdot \frac{T}{T^2} \geq 0 \qquad (5.57)$$

Equation (5.57) is the simplified form of the well-known *Clausius–Duhem inequality* (Acharya and Shawki 1996; Clarebrough et al. 1957; Halford 1966; Lemaitre and Chaboche 1990), which in the complete form reads

$$\dot{\gamma} = \frac{\sigma : \dot{\varepsilon}_p}{T} - J_q \cdot \frac{T}{T^2} - \frac{A_k \dot{V}_k}{T} \geq 0 \qquad (5.58)$$

---

[*] Note that the entropy production was symbolized by σ in previous chapters. To differentiate between σ used to denote stress and entropy production, we use $\dot{\gamma}$ to refer to the entropy production, hereafter.

where $\dot{V}_k$ can be any internal variables such as damage and hardening, and A_k are thermodynamic forces associated with the internal variables. In the derivation of the Clausius–Duhem inequality, it is assumed that variables A_k associated with the internal variables are defined by the specification of the thermodynamic potential ψ as

$$A_k = \rho \frac{\partial \psi}{\partial V_k} \tag{5.59}$$

For metals, the dissipation associated with evolution of internal variables $A_k \dot{V}_k$ represents only 5 to 10% of the entropy generation compared to plastic dissipation $\sigma : \dot{\varepsilon}_p$ and is often negligible (Acharya and Shawki 1996; Clarebrough et al. 1955, 1957; Ital'yantsev 1984a,b; Naderi, Amiri, and Khonsari 2010). Hence, it is commonly assumed that:

$$\frac{A_k \dot{V}_k}{T} \approx 0 \tag{5.60}$$

So, at least for metals, Equation (5.57) provides a reasonable approximation for calculation of the entropy generation in a system subjected to fatigue loading.

5.4 THERMODYNAMIC FORCES AND FLOWS IN PROCESSES INVOLVING FATIGUE

As discussed in Chapter 3, both thermodynamic forces and flows vanish when the system is in equilibrium state, that is, $X_k = 0$ and $J_k = 0$, and the entropy generation is zero. This implies that thermodynamic forces and flows exist under nonequilibrium or near-equilibrium conditions. In the analysis of entropy production in this chapter, we rely on the assumption of local-equilibrium condition. While at any small cell in the system, the assumption of thermodynamic equilibrium is valid and driving forces are nil, on the macroscopic scale the system could be out of equilibrium state.

Let us rewrite the entropy generation $\dot{\gamma}$ in terms of thermodynamic forces and flows. Recall Equation (3.7), which states

$$\frac{d_i S}{dt} = \sum_k X_k J_k \tag{5.61}$$

The assumption of local-equilibrium allows us to deal with the volumetric entropy generation $\dot{\gamma}$ instead of $d_i S/dt$. Starting with the general Clausius–Duhem inequality, Equation (5.58), we now rewrite it in terms of thermodynamic forces and flows as follows

$$\dot{\gamma} = \left(\frac{\sigma}{T}\right) : \dot{\varepsilon}_p - \left(\frac{T}{T^2}\right) \cdot J_q - \left(\frac{A_k}{T}\right) \dot{V}_k \geq 0 \tag{5.62}$$

Equation (5.62) can be interpreted as the product of thermodynamic forces:

$$X = \left\{\frac{\sigma}{T}, -\frac{T}{T^2}, -\frac{A_k}{T}\right\} \tag{5.63}$$

and generalized thermodynamic flows:

$$J = \{\dot{\varepsilon}_p, J_q, \dot{V}_k\} \tag{5.64}$$

Note that the thermodynamic force associated with heat flow, J_q, can be written as

$$X_q = -\frac{T}{T^2} = \left(\frac{1}{T}\right) \tag{5.65}$$

To construct a correlation between thermodynamic fluxes and forces, we start from the definition of dissipation potential function:

$$\Phi = T\,\dot{\gamma} \tag{5.66}$$

Using the expression for $\dot{\gamma}$ from Equation (5.57), the potential function Φ can be written as

$$\Phi = \sigma : \dot{\varepsilon}_p - J_q \cdot \frac{T}{T} \tag{5.67}$$

Equation (5.67) suggests that the dissipation potential can be expressed as a function of the thermodynamic fluxes $\Phi = \Phi(\dot{\varepsilon}_p,\ J_q)$. The second law of thermodynamics asserts that the dissipation potential is a positive convex function of the fluxes that vanishes at the origin of the space of the fluxes. It means that for zero fluxes ($\dot{\varepsilon}_p = J_q = 0$), the dissipation potential is nil, that is, $\Phi = 0$. Figure 5.9 shows a schematic of the dissipation potential function Φ plotted in the space of the flux variables when there are two fluxes, J_1 and J_2.

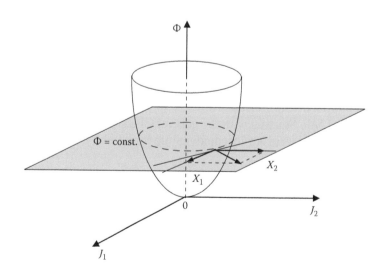

FIGURE 5.9 Schematic of the dissipation potential function plotted as a function of fluxes.

5.4.1 Legendre–Fenchel Transformation

Due to the convexity of the dissipation function, the Legendre–Fenchel transformation provides a systematic way of determining the thermodynamic forces associated with the fluxes. Presently, we begin by introducing the Legendre–Fenchel transformation and then proceed to apply it to the dissipation function $\Phi = \Phi(\dot{\varepsilon}_p, J_q)$. Readers interested in details of Legendre-Fenchel transformation may refer to Rockafellar (1970).

Let $\tilde{J}_1, \tilde{J}_2, \cdots, \tilde{J}_m \in \mathfrak{R}^m$ be parameters in domain $\mathfrak{R}^m$, and D be a convex function of $\tilde{J}_i$. Then there exist dual variables $\tilde{X}_1, \tilde{X}_2, \cdots, \tilde{X}_m \in \mathfrak{R}^m$ which can be defined as

$$\tilde{X}_i = \frac{\partial D}{\partial \tilde{J}_i} \tag{5.68}$$

Equation (5.68) is useful for determination of the dual variables, $\tilde{X}_m$ when the convex function D is defined in $\mathfrak{R}^m$ with parameters $\tilde{J}_m$ in its arguments. However, for the function D, a dual function $D*$ can be defined in $\mathfrak{R}^m$ with $\tilde{X}_m$ as its arguments. The dual function $D*$ is the transformation of D from parameters space $\tilde{J}_m$ to dual variable space $\tilde{X}_m$. The dual function $D*$ of D from Legendre–Fenchel transformation is

$$D*\left(\tilde{X}_1, \tilde{X}_2, \cdots, \tilde{X}_m\right) = \operatorname*{Sup}_{\left(\tilde{J}_1, \tilde{J}_2, \cdots, \tilde{J}_m\right)} \left[\sum_i \tilde{J}_i \tilde{X}_i - D\left(\tilde{J}_1, \tilde{J}_2, \cdots, \tilde{J}_m\right) \right] \tag{5.69}$$

Also, it follows from Equation (5.69) that:

$$\tilde{J}_i = \frac{\partial D*}{\partial \tilde{X}_i} \tag{5.70}$$

The transformation in Equation (5.69) conserves convexity (Rockafellar 1970). Thus, the dual (conjugate) function $D*$ is a convex function of dual variables $\tilde{X}_i$.

By applying the Legendre–Fenchel transformation to the potential function Φ, we can introduce a dual (conjugate) potential function $\Phi*$ written as

$$\Phi*\left(\sigma, - \ T/T\right) = \operatorname*{Sup}_{(\sigma, - \ T/T)} \left[\sigma : \dot{\varepsilon}_p - J_q \cdot \frac{T}{T} - \Phi(\dot{\varepsilon}_p, J_q) \right] \tag{5.71}$$

Note that the conjugate potential $\Phi*$ is a convex function of forces (dual variables) σ and J_q. Equations (5.68) and (5.70) suggest the following definitions of thermodynamic forces and fluxes:

$$\sigma = \frac{\partial \Phi}{\partial \dot{\varepsilon}_p}$$

$$-\frac{T}{T} = \frac{\partial \Phi}{\partial J_q} \tag{5.72}$$

and

$$\dot{\varepsilon}_p = \frac{\partial \Phi *}{\partial \sigma}$$

$$J_q = \frac{\partial \Phi *}{\partial \left(- \ T/T\right)} \tag{5.73}$$

Equation (5.72) suggests that σ and $\left(- \ T/T\right)$ are the components of the $\nabla\Phi$ in the space of the thermodynamic fluxes, $\dot{\varepsilon}_p$ and J_q. This is schematically shown in Figure 5.9, wherein thermodynamic forces, X_1 and X_2, are the components of $\nabla\Phi$ normal to the Φ = constant.

Let us make use of the characteristics of the conjugate potential function as expressed by Equation (5.73). Taking the derivative of $\dot{\varepsilon}_p$ with respect to $\left(- \ T/T\right)$ results in

$$\frac{\partial \dot{\varepsilon}_p}{\partial(- \ T/T)} = \frac{\partial}{\partial(- \ T/T)}\left(\frac{\partial \Phi *}{\partial \sigma}\right) = \frac{\partial}{\partial \sigma}\left(\frac{\partial \Phi *}{\partial\left(- \ T/T\right)}\right) = \frac{\partial J_q}{\partial \sigma} \tag{5.74}$$

The relationship between the derivative of the thermodynamic forces and fluxes in Equation (5.74) is the so-called nonlinear Onsager reciprocal relation, which, in general form, states that the derivative of flux J_i with respect to force X_j equals to the derivative of flux J_j with respect to force X_i:

$$\frac{\partial J_i}{\partial X_j} = \frac{\partial J_j}{\partial X_i} \quad \forall \ i \neq j \tag{5.75}$$

In a special case, if the dissipation potential function Φ takes a quadratic form, the fluxes become linear function of forces, $J_i = L_{ij}X_j$, and the well-established reciprocal Onsager relation between phenomenological coefficients L_{ij} is reproduced as

$$L_{ij} = L_{ji} \quad \forall \ i \neq j \tag{5.76}$$

Equation (5.76) is established for the state of near-equilibrium where fluxes and forces both deviate by a small amount from equilibrium. Recall from Section 3.2 that for some irreversible processes such as plastic deformation of solid materials, the assumption of linear relation between forces and flows is invalid and consideration of the Onsager's reciprocal relation is, in fact, inapt (Ziegler 1983).

REFERENCES

Acharya, A. and Shawki, T.G. 1996. The Clausius-Duhem inequality and the structure of rate-independent plasticity. *Int. J. Plasticity* 12, 229-238.
Aflatooni, K., Hornsey, R., and Nathan, A. 1997. Thermodynamic treatment of mechanical stress gradient in coupled electro-thermo-mechanical systems. *Sensors Mater.* 9, 449–456.
Al Nassaar, Y.N. 2003. Convective heating of solid surface: Entropy generation due to temperature field and thermal displacement. *Entropy* 5, 467–481.

Amiri, M., Naderi, M., and Khonsari, M.M. 2011. An experimental approach to evaluate the critical damage. *Int. J. Damage Mech.* 20, 89–112.

Atkinson, C. 1991. Exact solution of a model problem for the coupled thermoelastic equations with nonsmall temperature changes. *J. Thermal Stresses* 14, 215–225.

Basaran, C. and Yan, C.Y. 1998. A thermodynamic framework for damage mechanics of solder joints. *J. Electronic Packaging* 120, 379–384.

Beghi, M. 1982. Measurement of the entropy production due to irreversible deformation of metals. *Il Nuovo Cimento* 1D, 778–788.

Biot, M.A. 1955. Variational principles in irreversible thermodynamics with application to viscoelasticity. *Physical Review* 97, 1463–1969.

Biot, M.A. 1956. Thermoelasticity and irreversible thermodynamics. *J. Applied Physics* 27, 240–253.

Biot, M.A. 1957. New methods in heat flow analysis with application to flight structures. *J. Aeronautical Sciences* 24, 857–873.

Biot, M.A. 1958. Linear thermodynamics and the mechanics of solids. *Proc. Third U. S. National Congress Appl. Mech. Amer. Soc. Mech. Eng.* 1–18.

Biot, M.A. 1973. Non-linear thermoelasticity, irreversible thermodynamics and elastic instability. *Indiana University Math. J.* 23, 309–335.

Bishop, J.E. and Kinra, V.K. 1993. Thermoelastic damping of a laminated beam in flexure and extension. *J. Reinforced Plastics Composites* 12, 210–226.

Bishop, J.E. and Kinra, V.K.. 1997. Elastothermodynamic damping in laminated composites. *Int J Solids Structures* 34, 1075–1092.

Clarebrough, L.M., Hargreaves, M.E., Head, A.K., and West, G.W. 1955. Energy stored during fatigue of copper. *Trans. Am. Inst. Mining Metall. Eng.* 203, 99–100.

Clarebrough, L.M., Hargreaves, M.E., West, G.W., and Head, A.K. 1957. The energy stored in fatigued metals. *Proc. R. Soc. A* 242, 160–166.

Coleman, B.D. and Mizel, V.J. 1964. Existence of caloric equations of state in thermodynamics. *J. Chem. Phys.* 40, 1116–1125.

Crowe, D. and Feinberg, A. 2001. *Design for Reliability.* Boca Raton, FL: CRC Press.

De Groot, S.R. and Mazur, P. 1962. *Non-Equilibrium Thermodynamics*. New York: Interscience Publishers.

Dillon Jr., O. W. 1963. Coupled thermoplasticity. *J. Mech. Phys. Solids* 11, 21–33.

Giancane, S., Chrysochoos, A., Dattoma, V., and Wattrisse, B. 2009. Deformation and dissipated energies for high cycle fatigue of 2024-T3 aluminum alloy. *Theoretical and Applied Fracture Mech.* 52, 117–121.

Hackl, K., Fischer, F.D., and Svoboda., J. 2010. A study on the principle of maximum dissipation for coupled and non-coupled non-isothermal processes in materials. *Proc. R. Soc. A*, 467, 1186–1196.

Halford, G.R. 1966, The Energy Required for Fatigue. *ASTM J. Mater.* 1, 3–18.

Ital'yantsev, Y.F. 1984a. Thermodynamic state of deformed solids, Report 1, Determination of local function of state. *Strength Mater.* 16, 238–241.

Ital'yantsev, Y.F. 1984b. Thermodynamic state of deformed solids, Report 2, Entropy failure criteria and their application for simple tensile loading problems. *Strength Mater.* 16, 242–247.

Keenan, J.H. 1951. Availability and irreversibility in thermodynamics. *British J. Appl. Phys.* 2, 183–192.

Kinra, V.K. and Milligan. K.B. 1994. A second-law analysis of thermoelastic damping. *J. Appl. Mech.* 61, 71–76.

Kluitenberg, G.A. 1962. Thermodynamical theory of elasticity and plasticity. *Physica* 28, 217–232.

Lemaitre, J. and Chaboche, J.L. 1990. *Mechanics of Solid Materials.* Cambridge: Cambridge University Press,.

Naderi, M., Amiri, M., and Khonsari, M.M. 2010. On the thermodynamic entropy of fatigue fracture. *Proc. R. Soc. A* 466, 423–438.

Parkus, H. 1968. *Thermoelasticity*. Waltham, MA: Blaisdell Publication Company.

Pitarresi, G. and Patterson E.A. 2003. A review of the general theory of thermoelastic stress analysis. *J. Strain Anal.* 38, 405–417.

Rockafellar R.T. 1970. *Convex Analysis*. Princeton, NJ: Princeton University Press.

Sonntag, R.E., Borgnakke, C., and Van Wylen, G. J. 2003. *Fundamentals of Thermodynamics*. New York: John Wiley Sons, Inc.

Vengallatore, S. 2005. Analysis if thermoelastic damping in laminated composite micromechanical beam resonators. *J. Micromech. Microeng.* 15, 2398–2404.

Ziegler, H. 1983. *An Introduction to Thermomechanics*. Amsterdam: North-Holland Publishing Company.

6 Thermodynamic Assessment of Fatigue Failure

In this chapter, we show how the thermodynamic relations developed in the preceding chapters can be applied to determine the onset of fatigue failure. Specifically, we focus our attention to evaluating entropy generation and determining the fatigue life of the metals. Several illustrative examples are presented that bring to light the advantages of using the principles of thermodynamics over conventional approaches. For example, the prediction of fatigue life under variable amplitudes has always been complicated and no universally accepted treatment is available. To this end, the role of thermodynamic entropy offers a path forward.

6.1 LIMITATION OF CONVENTIONAL METHODS AND THE NEED FOR FURTHER ADVANCES

The conventional predictive models of fatigue failure are established based upon the examination of experimental data, typically in a laboratory setting. For example, the limitations of conventional S-N diagrams in predicting fatigue life become all the more evident when one seeks to apply the results to realistic environmental conditions and/or complex multi-axial loading, where no models can claim to be complete. It is, therefore, highly desirable to remove the major shortcomings associated with the conventional approaches of fatigue life prediction. In this chapter, we discuss how the concept of entropy can be used to quantify degradation associated with fatigue in a more general way.

6.2 EVALUATION OF ENTROPY GENERATION AND ENTROPY FLOW

Referring to Section 5.3, the relationship between the volumetric entropy generation, $\dot{\gamma}$, plastic energy dissipation, temperature gradient, and internal variables is

$$\dot{\gamma} = \frac{\sigma : \dot{\varepsilon}_p}{T} - J_q \cdot \frac{T}{T^2} + \frac{A_k \dot{V}_k}{T} \tag{6.1}$$

Considering no mass transfer, the entropy exchange with the surroundings, that is, the entropy flow per unit area, $d_e s/dt$, is related to the convective heat transfer (see Figure 6.1):

$$\frac{d_e s}{dt} = \frac{h(T - T_0)}{T} \tag{6.2}$$

The entropy flow is a function of heat transfer coefficient, h; the surface temperature of the specimen, T; and the surroundings temperature, T_0.

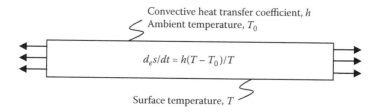

FIGURE 6.1 Entropy exchange with surroundings.

While surface temperature can be readily measured, the difficulty in evaluation of Equation (6.2) arises from the unknown convective heat transfer coefficient, h. The heat transfer coefficient is generally dependent on the thermal condition of the environment and the geometry of the specimen. For example, the heat transfer coefficient of a flat specimen under uniaxial load that is oriented vertically in a quiescent air can be readily evaluated from available empirical equation for natural convection heat transfer. However, for the fatigue tests in which the specimen does not remain stationary—such as in cyclic bending—the evaluation of convective heat transfer becomes more complicated.

To evaluate the entropy generation and the entropy flow, one must first determine the variation of surface temperature with time. This can be obtained either experimentally or by solving the appropriate governing equations as a function of time. The following example illustrates the procedure for evaluation of entropy when temperature history is available.

Example 6.1

Consider a bending plate specimen made of Aluminum 6061-T6 shown in Figure 6.2. The specimen is initially at room temperature, $T_0 = 28°C$. One end of the specimen is clamped and the other end is actuated cyclically, bending the specimen back and forth at the frequency of $f = 10$ Hz. The surface temperature of the plate at the location close to the clamped end, where the specimen is susceptible to fracture, is captured by means of an infrared (IR) camera and the result is shown

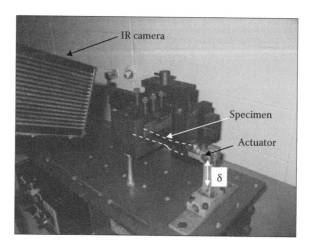

FIGURE 6.2 Specimen undergoing bending fatigue load.

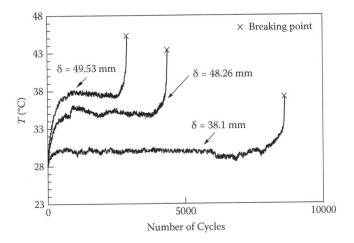

FIGURE 6.3 Temperature evolution for three different displacement amplitudes.

in Figure 6.3. This figure shows the temperature of three different tests at different displacement amplitudes, δ, representing different stress levels. Testing conditions are given in Table 6.1 (Amiri and Khonsari 2010). In this table, Δw denotes the rate of heat generation per cycle (see Chapter 4 for evaluation method).

(a) Estimate the entropy generation associated with the plastic deformation.
(b) Determine the accumulated entropy generation over the entire fatigue life and compare the entropy accumulation for the three different displacement amplitudes.

SOLUTION

(a) Assuming that the plastic deformation dominates the energy generation, we evaluate the entropy generation due to plastic deformation by merely considering the first term on the right-hand side of Equation (6.1)

$$\dot{\gamma} = \frac{\sigma : \dot{\varepsilon}_p}{T} \tag{6.3}$$

Noting that $\sigma : \dot{\varepsilon}_p = f\ w$, the entropy generation becomes

$$\dot{\gamma} = \frac{f\ w}{T} \tag{6.4}$$

TABLE 6.1
Testing Conditions

δ (mm)	N_f	Δw (kJ/m³cycle)
38.1	8610	148.7
48.26	4370	285.9
49.53	2910	423

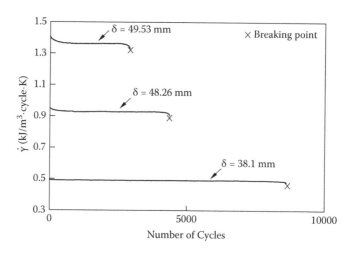

FIGURE 6.4 Evolution of entropy generation for three different displacement amplitudes

Figure 6.4 shows the evolution of the entropy generation for three different displacement amplitudes. Note that the temperature used for evaluation of the entropy generation is in Kelvin. It can be seen that for most of the fatigue life, the entropy generation remains nearly constant. For higher displacement amplitudes, the entropy generation is greater due to the greater heat generation.

(b) Equation (6.4) gives the rate of entropy generation. However, to find the accumulated entropy generation, γ_f, we need to integrate Equation (6.4) over time, from $t = 0$ to time $t = t_f$ when fracture occurs

$$\gamma_f = \int_{t=0}^{t=t_f} f\,\dot\gamma\,dt = \int_{t=0}^{t=t_f} f\,\frac{w}{T}\,dt \qquad (6.5)$$

Evaluating γ_f for three different testing conditions, we obtain

for $\delta = 49.53$ mm: $\gamma_f = 3.97$ (MJ/m³K);

for $\delta = 48.26$ mm: $\gamma_f = 4.05$ (MJ/m³K);

for $\delta = 38.1$ mm: $\gamma_f = 4.21$ (MJ/m³K).

An interesting observation is that values obtained for the accumulated entropy generation are nearly constant, averaging to $\gamma_f = 4.07$ (MJ/m³K), regardless of the displacement amplitude. This implies that the specimen fractures once the accumulation of entropy generated reaches a certain value. The importance of this finding can be illustrated by plotting the evolution of the accumulation of the entropy generation against the number of cycles for the three testing conditions. Figure 6.5 shows the variation of the entropy as it accumulates over the number of cycles. At the beginning of the test, the accumulation of entropy is nil and it linearly increases until it reaches roughly 4.07 (MJ/m³K), at which point fracture occurs. ▲

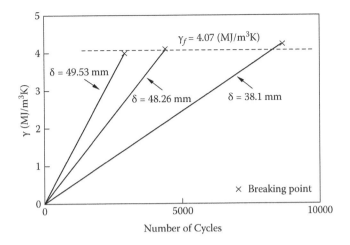

FIGURE 6.5   Evolution of accumulation of entropy generation for Aluminum 6061-T6.

Example 6.1 illustrates an interesting characteristic of fatigue failure of metals, which can be generalized to a broader set of testing conditions. Researchers have postulated and experimentally demonstrated that metals subjected to low-cycle fatigue load fracture upon reaching a fixed value of accumulated entropy regardless of the testing conditions. In fact, the entropy accumulation during a fatigue test is found to be independent of the frequency, the load amplitude, the size and the geometry of the specimen, and the type of loading. A recent paper by Naderi, Amiri, and Khonsari (2010) shows that entropy gain up to failure can be viewed as a property of the material. The notion of constant entropy gain at fatigue failure offers a methodology for prediction and prevention of the fatigue failure. This is illustrated in the following section.

6.3   TIME TO FAILURE

The time at which fatigue failure occurs can be defined in different ways. Some researchers define it as the onset of the appearance of small detectable cracks or the time upon which a crack grows to a critical length. Others define it as the onset of a change in one or more of the material's properties such as Young's modulus, electrical resistance, and thermal conductivity. Ital'yantsev (1984a, 1984b) postulates that the fatigue failure occurs when the accumulation of entropy reaches a critical value. To use this concept, the increase in the entropy during cyclic loading as well as the *initial entropy* prior to fatigue must be known. Clearly, it is impractical to measure the initial entropy of a material. Ital'yantsev also proposes mathematical conditions for failure based on the entropy generation (not the total entropy). According to Ital'yantsev, the entropy of a material tends to increase until failure occurs, after which the entropy production is vanished

$$\text{at } t = 0, \; d_i s/dt = 0 \tag{6.6a}$$

$$\text{for } t > 0, \; d_i s/dt > 0 \tag{6.6b}$$

$$\text{at } t = t_f, \; d_i s/dt = 0 \tag{6.6c}$$

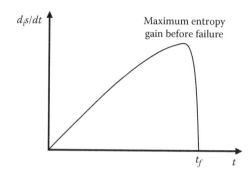

FIGURE 6.6 Entropy generation increases during fatigue until reaching maximum before the final fracture.

These conditions satisfy the second law of thermodynamics and necessitate the existence of a maximum entropy generation before failure. This can be viewed as the maximum entropy generation barrier. The terminology used here represents a deterministic amount of entropy generation that must be surmounted to cause failure. Figure 6.6 schematically shows the conditions presented in Equations (6.6).

6.3.1 FAILURE CRITERION BASED ON ACCUMULATION OF ENTROPY GENERATION

A similar concept is presented in the extensive studies of Whaley (1983a, 1983b, 1983c, 1984), Whaley, Pao, and Lin (1983), and Whaley, Chen, and Smith (1983) where they postulate that the entropy gain during fatigue is related to the plastic energy dissipation and can be estimated by integrating the plastic energy per temperature of material. It is hypothesized that the total entropy gain due to irreversible plastic deformation is a material constant at the onset of fatigue failure. By integrating Equation (6.1) from time $t = 0$ to fracture time $t = t_f$ and neglecting the effect of internal variables on entropy generation, we arrive at the following expression for the total entropy gain, γ_f:

$$\gamma_f = \int_0^{t_f} (\sigma : \dot{\varepsilon}_p / T - \boldsymbol{J}_q \cdot \ T/T^2)\, dt \qquad (6.7)$$

The notion that the total entropy gain γ_f at the onset of fatigue failure is a material constant has been hypothesized in many papers. However, determination of γ_f for a specific material has only recently become available. Naderi, Amiri, and Khonsari (2010) evaluate the entropy accumulation in a series of fatigue experiments conducted using Stainless Steel 304L and Aluminum 6061-T6. They employ Equation (6.7) and show that within the range of their experiments, the entropy generation associated with the plastic deformation is the dominant term in Equation (6.7). They further verify that the fatigue failure entropy (FFE) is a constant, regardless of the type of loading (torsion, axial, bending), frequency of the test, size of the specimen, and loading amplitude. The reported value for Stainless Steel 304L is about $\gamma_f = 60$ (MJ/m³K) and for Aluminum 6061-T6 is about $\gamma_f = 4$ (MJ/m³K). In the following example, we borrow experimental results from the work of Naderi, Amiri,

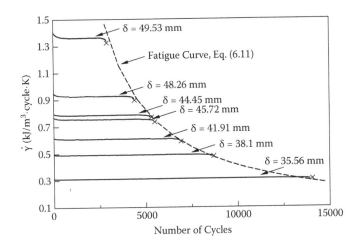

FIGURE 6.7 Fatigue curve predicted based on the concept of entropy generation.

and Khonsari (2010) to put forward a criterion for prediction of the failure time based on the concept of constant FFE.

Example 6.2

Figure 6.7 shows the results of a series of bending fatigue tests where entropy generation is plotted against number of cycles. The results pertain to fatigue tests of Aluminum 6061-T6 at different displacement amplitudes. Use Equation (6.7) with plastic deformation as the dominant source of entropy generation to arrive at the criterion for prediction of failure time, or equivalently number of cycles to failure.

SOLUTION

Assuming that the heat conduction is negligible, Equation (6.7) reduces to Equation (6.5) as in Example 6.1. Therefore, the entropy accumulation is simply:

$$\gamma_f = \int_{t=0}^{t=t_f} f\,\dot{\gamma}\,dt \tag{6.8}$$

Equation (6.8) indicates that the γ_f is the area underneath each curve shown in Figure 6.7. The hypothesis of constant FFE implies that the areas covered underneath each curve associated with the given displacement amplitude, δ, must be equal. Therefore, if $(\dot{\gamma})_1$, $(\dot{\gamma})_2$,... $(\dot{\gamma})_n$ denote the entropy generations at stress levels δ_1, δ_2, ..., δ_n, respectively, we have

$$\gamma_f = \int_{t=0}^{t=(t_f)_1} (\dot{\gamma})_1\,dt = \int_{t=0}^{t=(t_f)_2} (\dot{\gamma})_2\,dt = ... = \int_{t=0}^{t=(t_f)_n} (\dot{\gamma})_n\,dt \tag{6.9}$$

where $(t_f)_1$, $(t_f)_2$, ..., $(t_f)_n$ are the times to failure at each stress level δ_1, δ_2, ..., δ_n, respectively. Figure 6.7 shows that the entropy generation, $\dot{\gamma}$, at a stress level set by the specified actuation displacement, δ, remains almost constant during the entire fatigue life. Considering this, Equation (6.9) reduces to

$$\gamma_f = (\dot{\gamma})_1(N_f)_1 = (\dot{\gamma})_2(N_f)_2 = \cdots = (\dot{\gamma})_n(N_f)_n \qquad (6.10)$$

Note that in Equation (6.10) the time to failure, t_f, is replaced by the number of cycles to failure, N_f.

Equation (6.10) suggests a new criterion for fatigue failure based on the concept of constant entropy accumulation at the onset of failure. Solving for $(N_f)_i$ yields

$$(N_f)_i = \frac{\gamma_f}{(\dot{\gamma})_i}, \quad i = 1, 2, \ldots, n \qquad (6.11)$$

The average value of FFE for the tests shown in Figure 6.7 is $\gamma_f = 4.1$ (MJ/m³K). Using this value, the number of cycles to failure at each stress level can be predicted from Equation (6.11). The fatigue curve as predicted by Equation (6.11) shows good agreement with the experimental results of Naderi, Amiri, and Khonsari (2010).▲

6.3.2 COFFIN–MANSON EQUATION AND FFE

It is interesting to mention that empirical correlations for fatigue analysis such as the Coffin–Manson equation and/or Miner's rule can be deduced from the consideration of the FFE concept. Of particular interest here is to arrive at the Coffin–Manson correlation beginning with the FFE concept. Discussion on Miner's rule and its thermodynamics derivation is given in Chapter 8. See Amiri and Khonsari (2012) for a full account of these analyses.

Briefly, the Coffin–Manson relationship presents a correlation between the fatigue life, N_f, and the plastic strain range, $\Delta\varepsilon_p/2$, as

$$N_f = C\left(\frac{\varepsilon_p}{2}\right)^n \qquad (6.12)$$

where C and n are empirical constants and are obtained from a series of low-cycle fatigue tests. In fact, the constant n represents the slope of the strain range versus the fatigue life curve when presented on the log-log plot.

To show how the FFE concept can be used to arrive at the Coffin–Manson equation, let us start with the definition of FFE given in Equation (6.7). Since we are dealing with low-cycle fatigue, the effect of heat conduction on entropy generation can be neglected and Equation (6.5) can be used. Therefore,

$$\gamma_f = \int_0^{t_f} \frac{f\ w}{T} dt \qquad (6.13)$$

Note that Δw represents the entropy generation by plastic deformation since it is the dominant source in low-cycle fatigue (see Naderi, Amiri, and Khonsari 2010). As discussed in Section 4.3, the plastic energy dissipation is approximately constant throughout the low-cycle fatigue test; therefore, Equation (6.13) can be written as

$$\gamma_f = \frac{w}{T} N_f \qquad (6.14)$$

In deriving Equation (6.14), it is also assumed that the temperature during fatigue is constant. This was discussed in Section 4.4 in relationship to the second phase of temperature evolution where steady state trait governs most of the fatigue life. Substituting the plastic energy term Δw with the expression in Equation (4.19) and rearranging the resulting equation yields

$$N_f = \frac{T\gamma_f(\varepsilon'_f)^{n'}}{4\sigma'_f\left(\dfrac{1-n'}{1+n'}\right)}\left(\frac{\varepsilon_p}{2}\right)^{-(1+n')} \tag{6.15}$$

which is in the form of Equation (6.12) with

$$\begin{cases} C = \dfrac{T\gamma_f(\varepsilon'_f)^{n'}}{4\sigma'_f\left(\dfrac{1-n'}{1+n'}\right)} \\ \\ n = -(1+n') \end{cases} \tag{6.16}$$

It is worth mentioning that the constant C in the Coffin–Manson equation is correlated to the FFE, γ_f, via Equation (6.16). Here, Equation (6.15) is obtained directly by entropy consideration without having to resort to an empirical model.

6.3.3 FAST PREDICTION OF FATIGUE FAILURE

Let us now turn our attention to another interesting result that emerges from the concept of constant entropy gain at fatigue failure. Figure 6.8 shows the normalized accumulation of entropy during the bending fatigue with Aluminum 6061-T6 for the test series presented in Example 6.2. The abscissa of Figure 6.8 shows the number of cycles normalized by dividing the final number of cycles when failure occurs. The ordinate shows the evolution of entropy accumulation normalized by dividing the entropy gain at the final failure, γ_f. It can be seen that the normalized entropy generation monotonically increases until it reaches the entropy at the failure point.

The relation between the normalized entropy accumulation and the normalized number of cycles is approximately linear and can be written as

$$\frac{\gamma}{\gamma_f} \cong \frac{N}{N_f} \tag{6.17}$$

Considering Equation (6.17), the number of cycles to failure, N_f, can be expressed as

$$N_f \cong \frac{1}{\gamma/N}\gamma_f \tag{6.18}$$

The value γ/N in Equation (6.18) represents the slope of the curve that describes the evolution of accumulated entropy. As demonstrated in Example 6.1 and also shown in Figure 6.5,

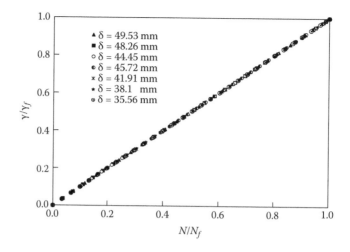

FIGURE 6.8 Normalized entropy accumulation versus normalized number of cycles. (Reproduced from Naderi, M., Amiri, M., and Khonsari, M. M., *Proc. R. Soc. A* 466, 426–438, 2010.)

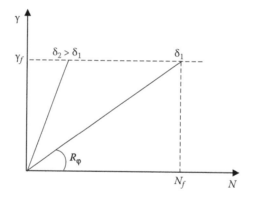

FIGURE 6.9 Schematic of evolution of entropy accumulation for two different stress levels.

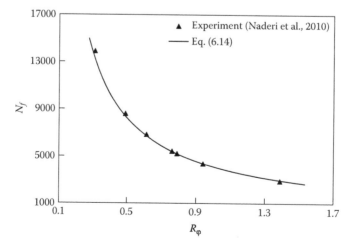

FIGURE 6.10 Fatigue life prediction based on the slope of the entropy accumulation.

the relation between γ and N is almost linear. This is schematically shown in Figure 6.9. For higher stress level (or higher displacement amplitude, δ), the slope of the entropy accumulation, R_φ, is higher.

Substituting γ/N by R_φ, Equation (6.18) yields

$$N_f \cong \frac{1}{R} \gamma_f \tag{6.19}$$

In Equation (6.19), the value of γ_f is known; therefore, to estimate the number of cycles to failure, N_f, one merely needs to evaluate the slope, R_φ, at the beginning of the test. Figure 6.10 shows the fatigue life prediction based on the model presented by Equation (6.19) for the experimental data of Example 6.2. The average value of FFE for Aluminum 6061-T6 is $\gamma_f = 4.1$ (MJ/m³K). The slope of the entropy accumulation is evaluated from the data at the very beginning of the test, that is, almost at 5% of the total fatigue life. For example, the slope of the entropy accumulation for displacement amplitude of $\delta = 35.56$ mm is evaluated to be $R_\varphi = 0.31$ (kJ/m³.cycle.K). Considering $\gamma_f = 4.1$ (MJ/m³K), the number of cycles to failure calculated from Equation (6.19) is $N_f = 4.1 \times 10^3/0.31 = 13,225$ cycles. The experimentally reported number of cycles to failure is $N_f = 13,900$ (about 4.9% error).

REFERENCES

Amiri, M. and Khonsari, M.M. 2010. Rapid determination of fatigue failure based on temperature evolution: Fully reversed bending load. *Int. J. Fatigue* 32, 382–389.

Amiri, M. and Khonsari, M.M. 2012. On the role of entropy generation in processes involving fatigue. *Entropy* 14, 24–31.

Ital'yantsev, Y. F. 1984a. Thermodynamic state of deformed solids. Report 1. Determination of local function of state. *Strength Mater.* 16, 238–241.

Ital'yantsev, Y. F. 1984b. Thermodynamic state of deformed solids. Report 2. Entropy failure criteria and their application for simple tensile loading problems. *Strength Mater.* 16, 242–247.

Naderi, M., Amiri, M., and Khonsari, M.M. 2010. On the thermodynamic entropy of fatigue fracture. *Proc. R. Soc. A* 466, 423–438.

Whaley, P.W. 1983a. A thermodynamic approach to metal fatigue. *Proc. ASME Int. Conf. Advances in Life Prediction Methods,* Albany, NY, 18–21.

Whaley, P.W. 1983b. A thermodynamic approach to material fatigue. *Proc. ASME Int. Conf. Advances in Life Prediction Methods.* Albany, NY, 41–50.

Whaley, P.W. 1983c. Entropy Production during Fatigue as a Criterion for Failure, The Critical Entropy Threshold: A Mathematical Model for Fatigue, *ONR Technical Report No.1, College of Engineering.* Lincoln: University of Nebraska.

Whaley, P.W. 1984. Entropy Production during Fatigue as a Criterion for Failure, A Local Theory of Fracture in Engineering Materials. *ONR Technical Report No.2, 1984, College of Engineering,* Lincoln: University of Nebraska.

Whaley, P.W., Chen, P.S., and Smith, G.M. 1983. Continuous measurement of material damping during fatigue tests. *Exp. Mech.* 24, 342–348.

Whaley, P.W., Pao, Y.C., and Lin, K.N. 1983. Numerical simulation of material fatigue by a thermodynamic approach. *Proc. 24th AIAA/ASME/ASCE/AHS Struct. Structural Dynamic and Materials Conference,* Lake Tahoe, Nevada, 544–551.

7 Damage Mechanics
An Entropic Approach

In this chapter, we briefly discuss how to implement the thermodynamic framework developed in preceding chapters to fracture mechanics. Particular attention is given to the application of the entropic approach in continuum damage mechanics (CDM). The damage variable, D, is first defined in the context of damage mechanics and is then linked to the notion of entropy flow and entropy generation. We demonstrate the practical usefulness of an entropy-based damage model in fatigue problems with variable loading amplitude and different stress states. The application of CDM-based models in fretting-fatigue and rolling-fatigue is also briefly presented.

7.1 INTRODUCTION TO DAMAGE MECHANICS

During past decades, a variety of damage models applicable to fatigue have been developed. Here, we do not intend to present an exhaustive literature review of available damage models. Rather, we present a brief review of papers that are relevant to our discussion.

Recall the application of the linear cumulative fatigue damage model known as Miner's rule (Equation 4.13) for assessment of variable-amplitude loading by counting the number of cycles. Let us rewrite Equation (4.13) in the following form:

$$D = \sum \frac{n_i(\sigma_i)}{N_i(\sigma_i)} \tag{7.1}$$

where n_i and N_i are the number of cycles and the number of cycles to failure at stress level σ_i, respectively. The damage variable, D, is a measure of loss in the material's strength against fatigue load. For a pristine material, $D = 0$, and it continuously increases as the fatigue damage progresses until failure occurs ($D = 1$). As mentioned in Chapter 4, Equation (7.1) falls short of addressing the loading sequence and modified versions of it have been developed to rectify the shortcomings of the original Miner's rule. For example, the so-called bi-linear (or double linear) damage model proposed by Grover (1960) accounts for loading sequence by treating the crack initiation and crack propagation stages differently. For crack initiation stage, the damage equation is defined by

$$D = \sum \frac{n_i}{\alpha N_i} \tag{7.2a}$$

and for crack propagation stage:

$$D = \sum \frac{n_i}{(1 - \alpha)N_i} \tag{7.2b}$$

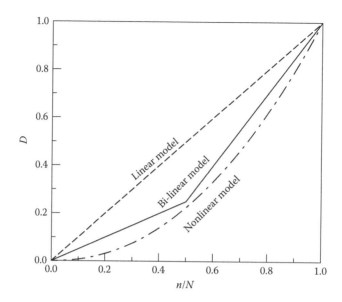

FIGURE 7.1   Damage variable versus life fraction.

where α is the fraction of life required for crack initiation. Having known the value of α for different stress levels, the bi-linear model provides good agreement with experimental results (Manson and Halford 1981). In fact, at low stress levels most of the life is expended for crack initiation, while at high stresses major fraction of life is spent for crack propagation (Stephens et al. 2000). This way, the bi-linear damage model provides a satisfactory prediction for low-to-high or high-to-low loading conditions. Figure 7.1 schematically shows the bi-linear damage model compared with linear Miner's rule.

Nonlinear damage models have also been developed to improve the Miner's rule and account for load sequence. For example, Marco and Starkey (1954) proposed a nonlinear damage model in the form of

$$D = \sum \left(\frac{n_i}{N_i} \right)^{\beta_i} \tag{7.3}$$

where β_i is the power exponent and varies as the stress level changes. Figure 7.1 shows a schematic of a nonlinear damage model along with the linear and bi-linear models. Note that efforts made to improve the Miner's rule have been only partially successful and no model can claim to work perfectly in a complex variable–amplitude loading. In fact, Stephens et al. (2000), stated, "Consequently, the Palmgren–Miner linear damage rule is still dominantly used in fatigue analysis or design in spite of its many shortcomings." A comprehensive review of available damage models is given by Fatemi and Yang (1998).

7.1.1   ENTROPY-BASED DAMAGE VARIABLE

The damage models reviewed in the previous section rely on counting the number of cycles, n_i, or more precisely the fraction of life, n_i/N_i. This provides an extremely easy technique for assessing the cumulative fatigue damage in practice since counting the cycles can be readily performed by using a counter. However, note that simply counting the cycles cannot account for parameters that significantly influence the damage accumulation such as alteration of the state of stress (axial, biaxial or multi-axial), variable frequency, and changes in environmental conditions (temperature, humidity, etc.). To overcome these shortcomings, energy-based damage models provide promising results by bringing into play the role of, for example, the hysteresis energy dissipation, Δw. A simple form of energy-based damage model can be defined by tallying up the energy dissipation per cycle as (Kliman 1984):

$$D = \sum \frac{w}{W_f} \tag{7.4}$$

where W_f is the energy at fracture. Utilizing the energy-based damage model, one can accumulate the hysteresis energy dissipated per cycle in each loading sequence. This way, the effect of different stress states, frequency, and so forth, is naturally taken into account.

In the sense of degradation and damage accumulation, the concept of tallying entropy is more fundamental than the energy dissipation. Recently, attempts have been made to link the damage variable, D, to entropy accumulation (Amiri, Naderi, and Khonsari 2011; Naderi and Khonsari 2010a, 2010b). Both entropy flow and entropy generation are proven to be good candidates for evaluation of damage.

Amiri, Naderi, and Khonsari (2011) reported a series of bending fatigue tests to evaluate damage parameters in which the entropy exchange with the surroundings, d_eS, was employed. As discussed in Chapter 2, the rate of entropy exchange with the surroundings can be evaluated from Equation (2.25). The accumulated entropy flow was computed by integrating Equation (2.25) over cycles (time) as

$$S_e = \int_0^t \frac{d_eS}{dt} dt = \int_0^t \frac{hA(T - T_0)}{T} dt \tag{7.5}$$

Note that if the integration is performed from the beginning, $t = 0$, to the failure, $t = t_f$, then the entropy evaluated from Equation (7.5) represents the total entropy flow at failure, S_f.

Figure 7.2 shows the normalized number of cycles as a function of normalized entropy exchange for different displacement amplitudes of bending fatigue test. Results are adapted from Amiri, Naderi, and Khonsari (2011). The material used for the specimen is Aluminum 6061-T6 and the frequency of tests is 10 Hz. The normalized number of cycles is defined as the ratio of the number of cycles to the number of cycles to failure, N/N_f, and the normalized entropy exchange is defined as the ratio of accumulation of entropy exchange to the entropy exchange at the failure, S_e/S_f. Results show a linear relationship between normalized number of cycles and normalized accumulation of entropy exchange, expressed as

$$\frac{S_e}{S_f} = \frac{N}{N_f} \tag{7.6}$$

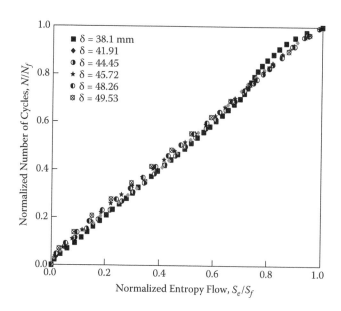

FIGURE 7.2 Normalized number of cycles against normalized entropy exchange. (Reproduced from Amiri, M., Naderi, M., and Khonsari, M.M., *Int. J. Damage Mech.*, 20, 89–112, 2011.)

Taking advantage of the above linear relationship, a new damage equation is defined based on the entropy exchange as follows (Amiri et al. 2011):

$$D = \frac{-1}{\ln\left(S_f\right)} \ln\left(1 - \frac{S_e}{S_f}\right) \tag{7.7}$$

Figure 7.3 illustrates the damage evolution for Stainless Steel 304 at displacement amplitude of $\delta = 48.26$ mm. Shown in this figure also are the results of the work of Duyi and Zhenlin (2001) for comparison. Duyi and Zhenlin employed the exhaustion of static fracture toughness to arrive at the following damage model:

$$D = \frac{1 - \sigma_a^2 / E U_{T0}}{\ln(N_f)} \ln\left(1 - \frac{N}{N_f}\right) \tag{7.8}$$

where U_{T0} is the static toughness of undamaged material, E is the elastic modulus, and σ_a is the stress amplitude. Figure 7.3 shows the nonlinear increase of the damage variable. Close to final failure, the damage variable increases drastically, representing the *critical damage*, D_c. This concept can be utilized as an indication of imminent fracture as discussed next.

Figure 7.4 shows the evolution of the damage parameter for Aluminum 6061-T6 undergoing a bending fatigue test at three different displacement amplitudes, δ. Amiri, Naderi, and Khonsari (2011) defined the critical condition in fatigue as a sudden increase in damage parameter which, in turn, is a consequence of a sharp increase in entropy exchange. This condition is shown in Figure 7.4 by a dashed line. Based on the entropy approach, they determined that the critical damage parameter for Aluminum 6061-T6 and Stainless Steel 304 are approximately 0.3 and 0.2, respectively. Critical damage is a material property and is independent of the loading amplitude.

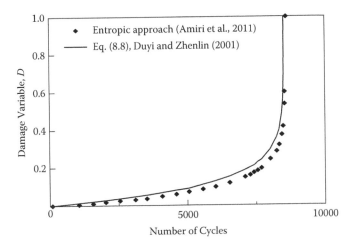

FIGURE 7.3 Evolution of D for Stainless Steel 304 at displacement amplitude of $\delta = 48.26$ mm. (Reproduced from Amiri, M., Naderi, M., and Khonsari, M.M., *Int. J. Damage Mech.*, 20, 89–112, 2011.)

Similarly, Naderi and Khonsari (2010a, 2010b) used the concept of entropy generation d_iS, instead, to derive a damage evolution model. Recall Equation (5.57) for entropy generation $\dot{\gamma}$ during the course of fatigue. The accumulated entropy generation can be evaluated by integrating Equation (5.57) over cycles (time) as

$$\gamma = \int_0^t \dot{\gamma}\,dt = \int_0^t \left(\frac{\boldsymbol{\sigma} : \dot{\boldsymbol{\varepsilon}}_p}{T} - \boldsymbol{J}_q \cdot \frac{T}{T^2} \right) dt \qquad (7.9)$$

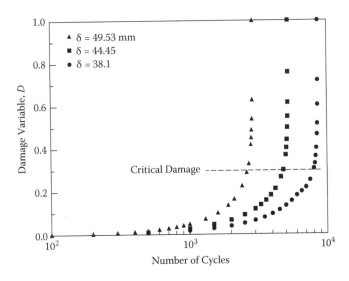

FIGURE 7.4 Critical damage for Aluminum 6061-T6. (Reproduced from Amiri, M., Naderi, M., and Khonsari, M.M., *Int. J. Damage Mech.*, 20, 89–112, 2011.)

Note that if the integration is performed from the beginning, $t = 0$, to the failure, $t = t_f$, entropy generation evaluated from Equation (7.9) represents the total entropy generation at failure, γ_f. Naderi and Khonsari used this definition to arrive at the following equation for the damage variable:

$$D = \frac{D_c}{\ln\left(1 - \gamma_c/\gamma_f\right)} \ln\left(1 - \frac{\gamma}{\gamma_f}\right) \tag{7.10}$$

where D_c denotes the critical value of damage and γ_c represents the entropy accumulation up to the critical condition. Figure 7.5 shows the evolution of the damage variable for two different loading sequences for Aluminum 6061-T6. Results are adapted from Naderi and Khonsari (2010a). Each sequence includes three loading stages. Plot (a) corresponds to the sequence from high to intermediate to low loads, while plot (b) is in reversed order. This figure clearly illustrates the capability of entropy generation in addressing the effect of load sequence on the damage variable. In fact, in test (a) where the load level increases, the damage induced in the sample is more pronounced than test (b). This is due to the fact that by applying higher load first, followed by lower load, greater damages in the form of macro-cracks is induced in the sample, which results in a higher value of D. This is in accordance with the concept of CDM, which is described next.

7.2   CONTINUUM DAMAGE MECHANICS (CDM)

Continuum damage mechanics is a relatively new branch of solid mechanics, which deals with analysis and characterization of a material's defect at micro- to mesoscale. The early studies date back to the works of Kachanov (1958) and major contributions were made later by Krajcinovic (1984), Lemaitre (1985), Kachanov (1986), and Chaboche (1988). Since then, CDM has been growing rapidly. Lemaitre (2002) presented the main landmarks of the development of the

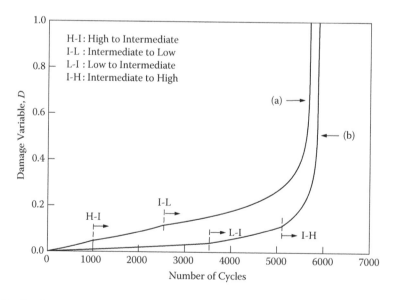

FIGURE 7.5 Evolution of D for variable-amplitude loading. (Reproduced from Naderi, M. and Khonsari, M.M., *J. Mater. Sci. Eng., A* 527, 6133–6139, 2010a.)

science of CDM from 1958 to 2000. In fact, the development of CDM-based damage models is rooted in irreversible thermodynamics analysis by taking into account the evolution of state variables, state potential, and dissipation potential (Lemaitre and Desmorat 2005). In what follows, we illustrate the application of the CDM model proposed by Bhattacharya (1997) and Bhattacharya and Ellingwood (1998, 1999) to predict crack initiation in fretting-fatigue and rolling-fatigue. The advantages of this approach are that this model is obtained from the laws of thermodynamics and that it uses the bulk material properties to predict the crack initiation. This model avoids the use of empirical relations that exist in the literature on fretting- and rolling-fatigue. Before delving into formulations, let us introduce basic definitions essential in CDM modeling.

7.2.1 DAMAGE VARIABLE, $D(n)$

In the CDM context, the damage variable, $D(n)$, is generally described by a scalar, second order, and fourth order tensor on the elemental area A_0, with normal vector n. It manifests itself in the gradual loss of *effective cross-sectional area* $\bar{A}$ (see Figure 7.6) and is defined as

$$D(n) = \frac{A_0 - \bar{A}}{A_0} \tag{7.11}$$

For a pristine material with effective cross-sectional area $\bar{A} = A_0$, Equation (7.11) yields $D(n) = 0$. For the isotropic damage, wherein the damage variable is independent of the normal vector, $D(n)$ reduces to a scalar denoted by D. To simplify the formulations, let us assume that the damage variable is scalar. Similar to the definition for effective cross-sectional area, the notion of effective stress can be defined based on D

$$\bar{\sigma} = \frac{\sigma}{1 - D} \tag{7.12}$$

Bhattacharya (1997) used the concept of effective stress along with the Ramberg–Osgood law to derive the relationship between stress and strain ε,

$$\varepsilon = \frac{\bar{\sigma}}{E} + \left(\frac{\bar{\sigma}}{K}\right)^M \tag{7.13}$$

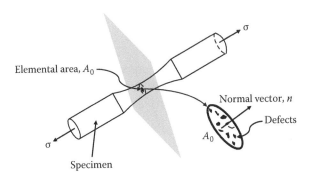

FIGURE 7.6   Definition of damage variable on elemental area A_0.

where E and K are elastic modulus and strain hardening modulus of pristine material, respectively, and M represents the hardening exponent. Note that the first term on the right-hand side of Equation (7.13) represents the elastic strain, ε_e, and the second term plastic strain, ε_p. According to Equation (7.13) the following relationship between ε_e and ε_p can be derived

$$\varepsilon_e = \frac{K}{E}\varepsilon_p^{\frac{1}{M}} \tag{7.14}$$

From Equation (7.12) through Equation (7.14), the effective stress can be written as

$$\bar{\sigma} = K(1-D)\varepsilon_p^{\frac{1}{M}} \tag{7.15}$$

Similar to the Equation (5.17), Bhattacharya (1997) defined the Helmholtz free energy as a function of the damage variable, $\psi = \psi(D)$, in which D is, in turn, a function of strain, $D = D(\varepsilon)$. He then arrived at the following equation for effective stress:

$$\bar{\sigma} = -\frac{d\psi}{d\varepsilon} = -\frac{\partial\psi}{\partial D}\frac{dD}{d\varepsilon} \tag{7.16}$$

Substituting Equation (7.15) into Equation (7.16) and rearranging the results, we obtain

$$\frac{dD}{d\varepsilon} = -\frac{K(1-D)\varepsilon_p^{1/M}}{\partial\psi/\partial D} \tag{7.17}$$

For the case of damage in uniaxial monotonic loading, Bhattacharya (1997) derived the following equation for $\partial\psi/\partial D$:

$$\frac{\partial\psi}{\partial D} = -\frac{K^2}{2E}\left(\varepsilon_p^{\frac{2}{M}} - \varepsilon_0^{\frac{2}{M}}\right) - \frac{K}{1+1/M}\left(\varepsilon_p^{1+\frac{1}{M}} - \varepsilon_0^{1+\frac{1}{M}}\right) - \frac{3}{4}\sigma_f \tag{7.18}$$

where ε_0 is the threshold plastic strain and σ_f is the fracture strength. Substitution of Equation (7.18) into Equation (7.17) yields

$$\frac{dD}{1-D} = \frac{K\varepsilon_p^{1/M}\,d\varepsilon}{\frac{K^2}{2E}\left(\varepsilon_p^{\frac{2}{M}} - \varepsilon_0^{\frac{2}{M}}\right) + \frac{K}{1+1/M}\left(\varepsilon_p^{1+\frac{1}{M}} - \varepsilon_0^{1+\frac{1}{M}}\right) + \frac{3}{4}\sigma_f} \tag{7.19}$$

The closed-form solution of the differential equation in Equation (7.19) can be obtained for uniaxial cyclic loading with D_i to be the damage variable at the ith number of the cycle. Without going through the mathematical derivations, let us present the final form of the equation for calculation of damage variable (Bhattacharya 1997):

$$D_i = \begin{cases} 1-(1-D_{i-1})F_i\ ; & \sigma_{max} \geq S_e \\ D_{i-1}\ ; & \sigma_{max} < S_e \end{cases} \tag{7.20a}$$

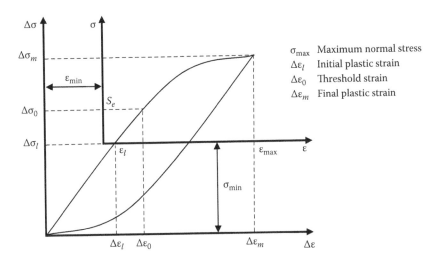

σ_max Maximum normal stress
Δε_l Initial plastic strain
Δε_0 Threshold strain
Δε_m Final plastic strain

FIGURE 7.7 Hysteresis loop and graphical definition of parameters in Equation (7.20b).

where

$$F_i = \frac{(1+1/M)^{-1}\ \varepsilon_{0i}^{1+1/M} \ - \ \varepsilon_{pli}^{1/M}\ \varepsilon_{0i} + C_i}{(1+1/M)^{-1}\ \varepsilon_{pmi}^{1+1/M} \ - \ \varepsilon_{pli}^{1/M}\ \varepsilon_{pmi} + C_i} \tag{7.20b}$$

and

$$C_i = \frac{3}{4}\frac{\sigma_f}{K} - \frac{\varepsilon_{0i}^{1+1/M}}{1+1/M} + \ \varepsilon_{pli}^{1/M}\ \varepsilon_{0i} \tag{7.20c}$$

In Equation (7.20a), σ_{max} is the maximum applied stress and S_e is the endurance limit. All other parameters in Equations (7.20b) and (7.20c) are graphically shown in Figure 7.7. If a fatigue test is performed under strain-controlled loading, $\Delta\varepsilon_{pmi}$, $\Delta\varepsilon_{pli}$, and C_i are independent of cycle i, and Equation (7.20a) reduces to the following form (Bhattacharya and Ellingwood 1999):

$$D_N = 1 - (1 - D_0)F^N \tag{7.21}$$

where N is the number of cycles and D_0 is the initial damage. For an initially undamaged material we have $D_0 = 0$. The CDM-based damage model presented by Equations (7.20) provides a methodology to analyze the damage evolution of damage in a wide range of processes. In fact, Bhattacharya (1997) and Bhattacharya and Ellingwood (1998, 1999) demonstrated the applicability of this model to predict ductile damage, creep damage, and fatigue damage. In what follows, we present some of the case studies that have utilized the CDM-based damage model to predict fatigue damage, fretting fatigue, and sliding wear.

7.2.2 CDM AND FATIGUE DAMAGE

Bhattacharya and Ellingwood (1999) applied the CDM approach presented by Equations (7.20) to predict the number of cycles to initiation of SAE 4340 Steel in a fully reversed

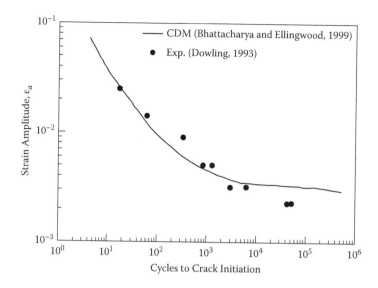

FIGURE 7.8 Prediction of fatigue damage for SAE 4340 steel. (Reproduced from Bhattacharya, B. and Ellingwood, B., *Int. J. Solids Struc.,* 36, 1757–1779, 1999.)

strain-controlled fatigue. Figure 7.8 shows the CDM prediction along with the experimental results of Dowling (1993) for comparison. This figure plots the strain amplitude, ε_a, against the number of cycles required to initiate a 0.038 mm crack. This crack length corresponds to the critical damage value of $D_c = 0.46$. The relatively good agreement of CDM prediction with experimental data shown in this figure demonstrates its capability in the prediction of fatigue damage.

7.2.3 CDM AND FRETTING FATIGUE

Damage due to repeated shear stress in oscillatory, small-amplitude motion known as fretting was discussed in Chapter 1. Quraishi, Khonsari, and Baek (2005) employed the CDM model presented in Equation (7.20) to study the crack initiation in fretting fatigue problems for two materials, namely Al 2024-T4 and S45C Carbon Steel. The subsurface alternating shear stresses were calculated and transformed to equivalent tension–compression stresses. Applying Equation (7.20), the number of cycles to initiate a fretting crack was calculated and compared to experimental data. Figure 7.9 shows the CDM prediction of fretting fatigue damage as compared to experimental work of Sato, Fuji, and Kodama (1986) for S45C. It is worth mentioning that the critical damage was assumed to be $D_c = 0.8$, in simulation of the results in this figure. Results of CDM prediction are in good agreement with experimental data, demonstrating the capability of CDM in predicting fretting fatigue. A more recent study that extends the results to predict fretting crack nucleation with provision for size effect is presented by Aghdam, Beheshti, and Khonsari (2012).

7.2.4 CDM AND SLIDING WEAR

Another study by Beheshti and Khonsari (2010) shows the application of the CDM model in prediction of wear under dry sliding contact. They applied Equations (7.20) to predict the

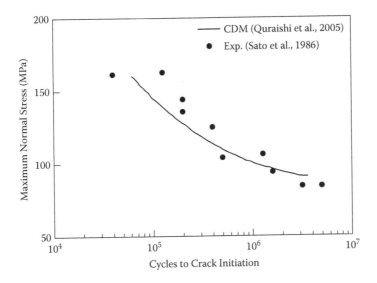

FIGURE 7.9 Prediction of fretting fatigue for S45C carbon steel. (Reproduced from Quraishi, S.M., Khonsari, M.M., and Baek, D.K., *Trib. Letters,* 19, 169–175, 2005.)

wear coefficient, k, and compared the simulation with experimental data of a series of pins on a disk wear test of Al 6061-T6. Wear coefficient is defined in Archard's law as

$$k = \frac{VH}{Nx} \tag{7.22}$$

where V is the wear volume, H denotes the hardness, N represents the normal load, and x is the sliding distance. The wear coefficient, k, is generally a material's property and independent of loading conditions. However, it depends on the surface roughness and the coefficient of friction. Figure 7.10 shows the CDM prediction of wear coefficient as a function

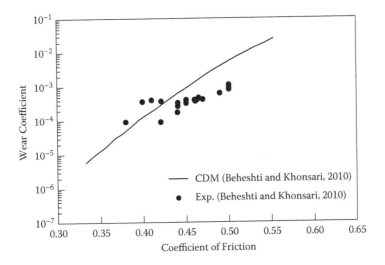

FIGURE 7.10 Prediction of wear coefficient for Al 6061-T6. (Reproduced from Beheshti, A. and Khonsari, M.M., *Trib. Int.,* 44, 1620–1628, 2011.)

of the coefficient of friction. The critical damage was assumed to be $D_c = 0.3$ in simulation of the results in this figure. The results of the CDM prediction of wear coefficient follow the experiments fairly well. A more recent study that extends the results to predict damage in rolling/sliding contacts with consideration of loading sequence has been carried out by Beheshti and Khonsari (2011).

REFERENCES

Aghdam, A.B., Beheshti, A., and Khonsari, M.M. 2012. On the fretting crack nucleation with provision for size effect. *Trib. Int.* 47, 32–43.

Amiri, M., Naderi, M., and Khonsari, M.M. 2011. An experimental approach to evaluate the critical damage. *Int. J. Damage Mech.* 20, 89–112.

Beheshti, A. and Khonsari, M.M. 2010. A thermodynamic approach for prediction of wear coefficient under unlubricated sliding condition. *Trib. Letters* 38, 347–354.

Beheshti, A. and Khonsari, M.M. 2011. On the prediction of fatigue crack initiation in rolling/sliding contacts with provision for loading sequence effect. *Trib. Int.* 44, 1620–1628.

Bhattacharya, B. 1997. A damage mechanics based approach to structural deterioration and reliability. Ph.D. thesis, Johns Hopkins University.

Bhattacharya, B. and Ellingwood, B. 1998. Continuum damage mechanics analysis of fatigue crack initiation. *Int. J. Fatigue* 20, 631–639.

Bhattacharya, B. and Ellingwood, B. 1999. A new CDM based approach to structural deterioration. *Int. J. Solids Struc.* 36, 1757–1779.

Chaboche, J.L. 1988a. Continuum damage mechanics. Part I—General concepts. *J. Appl. Mech.* 55, 59–64.

Chaboche, J.L. 1988b. Continuum damage mechanics. Part II—Damage growth, crack initiation, and crack growth. *J. Appl. Mech.* 55, 65–72.

Dowling, N. 1993. *Mechanical Behavior of Materials.* Englewood Cliffs, NJ: Prentice-Hall.

Duyi, Y. and Zhenlin, W. 2001. A new approach to low-cycle fatigue damage based on exhaustion of static toughness and dissipation of cyclic plastic strain energy during fatigue. *Int. J. Fatigue* 23, 679–687.

Fatemi, A. and Yang, L. 1998. Cumulative fatigue damage and life prediction theories: A survey of the state of the art for homogeneous materials. *Int. J Fatigue* 20, 9–34.

Gatts, R.R. 1961. Application of a cumulative damage concept to fatigue. *Trans. ASME, J. Basic Eng.* 83, 520–530.

Grover, H.J. 1960. An observation concerning the cycle ratio in cumulative damage. *Symp. Fatigue Aircraft Struc., ASTM STP 274*, Philadelphia. PA: *Am. Soc. Testing and Mater.* 20–124.

Kachanov, L.M. 1958. Time of the rupture process under creep conditions. [In Russian.] Izv Akad Nauk USSR. *Otd Tekh Nauk, OTN* 8, 26–31.

Krajcinovic, D. 1984. Continuum damage mechanics. *Appl. Mech. Rev.* 37, 1–6.

Kliman, V. 1984. Fatigue life prediction for a material under programmable loading using the cyclic stress-strain properties. *Mater. Sci. Eng.* 68, 1–10.

Kachanov, L.M. 1986. *Introduction to Continuous Damage Mechanics.* Dordrecht, The Netherlands: Martinus Nijhoff Publisher.

Lemaitre, J. 2002. Introduction to continuum damage mechanics. In *Continuum Damage Mechanics of Materials and Structures,* ed. O. Allix and F. Hild. Oxford, UK: Elsevier Science.

Lemaitre, J. 1985. A Continuous Damage Mechanics Model for Ductile Fracture. *J. Eng. Mater. Tech.* 107, 83–89.

Lemaitre, J. and Desmorat, R. 2005. *Engineering Damage Mechanics: Ductile, Creep, Fatigue and Brittle Failures.* Berlin and Heidelberg (Print. Netherlands): Springer-Verlag.

Manson, S.S. and Halford, G.R. 1981. Practical implementation of the double linear damage rule and damage curve approach for treating cumulative fatigue damage. *Int. J. Frac.* 17, 169–192.

Marco, S.M. and Starkey, W.L. 1954. A concept of fatigue damage. *Trans. ASME* 76, 627–632.

Naderi, M. and Khonsari, M.M. 2010a. A thermodynamic approach to fatigue damage accumulation under variable loading. *J. Mater. Sci. Eng. A* 527, 6133–6139.

Naderi, M. and Khonsari, M.M. 2010b. An experimental approach to low-cycle fatigue damage based on thermodynamic entropy. *Int. J. Solids and Struc.* 47, 875–880.

Quraishi, S.M., Khonsari, M.M., and Baek, D.K. 2005. A thermodynamic approach for predicting fretting fatigue life. *Trib. Letters* 19, 169–175.

Sato, K., Fuji, H., and Kodama, S. 1986. Crack propagation behavior in fretting fatigue of S45C carbon steel. *Bull. JSME* 29, 3253–3258.

Stephens, R.I., Fatemi, A., Stephens, R.R., and Fuchs, H.O. 2000. *Metal Fatigue in Engineering,* 2nd ed. New York: John Wiley & Sons, Inc.

8 Self-Organization in Fatigue

In preceding chapters one of the major concerns in fatigue failure analysis, that is, prediction of failure time, was discussed in detail, and we enumerated some of the prediction methods such as the stress-strain approach, energy approach, temperature approach, and thermodynamics approach. Other main concerns in fatigue analyses are prevention of failure and accelerated fatigue testing. In this chapter, the methodologies that improve fatigue life are discussed. Fatigue life extension methodologies can be broadly categorized into two groups: steps that can be taken at the manufacturing stage to improve the fatigue properties of the material and methods that can help to delay fatigue failure. In both categories, the notion of self-organization underlies a scientific base for analysis of the problem. The self-healing materials with healing agents, for example, belong to the first category, which is not the concern in this book. The second category, which is the main focus of this chapter deals with the investigation of the influence of external elements—electric current, magnetic field, environmental conditions, and so forth—that can retard time to fatigue failure and extend life. Here, our discussion pertains merely to metallic materials; nonmetals and composites are excluded.

8.1 INTRODUCTION TO SELF-ORGANIZATION

When a system reaches the equilibrium state, its entropy (and associate disorder) is at maximum. This implies that to increase the orderliness, the system should deviate from equilibrium state. According to Nicolis and Prigogine (1977), to drive the system further from equilibrium, the fluctuations from the average state should be above a certain critical value. For example, in a fatigue process, self-organization may occur above some critical density of the dislocations (Kabaldin and Murav'yev 2007). The source of fluctuation can be an external element(s) acting on the system to bring about deviation from the average state. Self-organization is directly associated with the mechanism of the formation of dissipative structures, which in a fatigue process of metals corresponds to the arrangement of new patterns of the material's microstructure formed during the structural transformation. More specifically, if we define structural transformation as the reconfiguration of microstructure during loading and unloading, then, the transformation can be examined by the forming and movement of partial dislocations and the accumulation of strain energy (Kabaldin and Murav'yev 2007).

The notion of self-organized dislocation has a rich history of supporting literature. Some of the pertinent work can be found in Kubin and Canova (1992); Glazov, Llantes, and Laird (1995); Suresh (1998); Kabaldin and Murav'yev (2007); Pisarenko, Voinalovich, and Mailo (2009); Taniguchi, Kaneko, and Hashimoto (2009); and Yang et al. (2011).

If during a fatigue process dissipative structures are formed, the system entropy will decrease and this consequently results in a slowing down of the fatigue damage. It is important to note, however, that the deviation of the system far from equilibrium is not the only necessary condition for the formation of dissipative structures. According to Prigogine's theorem (Nicolis and Prigogine 1977), the emergence of dissipative structures in a system obeying *linear* laws is impossible. As it was discussed in Chapter 3, according to Equation (3.15), each thermodynamic flux can depend on forces of all active processes in the system. However, if the dependency manifests itself in a linear manner, the possibility of formation of dissipative structures is unlikely. This implies that, for example, in a heat conduction process in which the heat flux is linearly dependent upon the gradient of temperature, it is theoretically impossible to attain ordered behavior. Similarly, in a purely elastic deformation of material (such as what is encountered in a very high-cycle fatigue), it is unlikely to promote self-organization. On the contrary, consider the low- and intermediate-cycle fatigue, where the plastic deformation is dominant and the fluxes do not depend linearly on forces. In this case, the irreversible process of plastic deformation may result in formation of different patterns in the material's structure with improved properties (e.g., hardening effect). Therefore, to develop self-organized behavior, we should take into account nonlinear, nonequilibrium characterization of the domain.

As mentioned earlier, dissipative structures are formed when the system deviates from average state. This can be induced externally via a magnetic field or electric current, or through environmental conditions, in order to drive the system far away from the equilibrium or the stationary state. The determination of the stability of the nonequilibrium stationary state can be performed based on Lyapunov's theory of stability (Glansdorff and Prigogine 1971). If a nonequilibrium stationary system loses its stability, the Lyapunov function $(\delta^2 S)/2$ should satisfy the following inequality:

$$\frac{1}{2}\delta^2 S < 0 \tag{8.1}$$

It can be shown that (Glansdorff and Prigogine 1971):

$$\frac{1}{2}\frac{d}{dt}(\delta^2 S) = \sum_k \delta X_k \delta J_k \tag{8.2}$$

where δX_k and δJ_k are deviations of thermodynamic forces and fluxes from stationary state. Equation (8.2) is called the *excess entropy generation*. If the excess entropy production is non-negative, the given state of the system is stable; otherwise, the system loses its stability. We now seek to show that through the application of the excess entropy production, a thermodynamic system, in general, and a fatigue system, in particular, can achieve self-organization.

Consider a system with two active dissipative processes. Application of Equation (8.2) yields

$$\frac{1}{2}\frac{d}{dt}(\delta^2 S) = \delta X_k \delta J_k = \delta X_1 \delta J_1 + \delta X_2 \delta J_2 \tag{8.3}$$

The variation of external elements can drive the system far from average state and contribute to entropy generation by the product of the forces with associate fluxes. Denoting the force and the flux of external element by X_2 and J_2, respectively, the entropy generation during a fatigue process can be described as

$$\frac{d_i S}{dt} = X_1 J_1 + X_2 J_2 \tag{8.4}$$

where $X_1 = \sigma/T$ and $J_1 = \dot{\varepsilon}_p$ denote the plastic deformation force and flows, respectively. The σ and $\dot{\varepsilon}_p$ denote the stress amplitude and the rate of plastic strain. The product of $X_2 J_2$ represents the contribution of the external element. If self-organization occurs, the entropy generation is minimal in a stationary state and the fatigue damage is retarded. Note that the flux of the external element, J_2, is the only variable which is controlled by the operator; that is, it can be the supplied electric current or the supplied magnetic field intensity of the rate of surface cooling. Assume that an electrical current, $J = I_e$, is supplied with voltage $X = V/T$. The contribution of this external element on the entropy generation is VI_e/T. The minimal entropy generation requires that

$$\frac{d}{dJ_2}\left(\frac{d_iS}{dt}\right) = \frac{d}{dJ_2}(X_1 J_1) + X_2 = 0 \tag{8.5}$$

Note that the external force, X_2, is assumed to be fixed and not influenced by the change in external flow, J_2. For example, one can supply a fixed voltage to the specimen under fatigue load while the electric current can change. Integrating Equation (8.5) yields

$$\int_{(X_1 J_1)_0}^{X_1 J_1} d(X_1 J_1) = -\int_{J_2=0}^{J_2} X_2 \, dJ_2 \tag{8.6}$$

where $(X_1 J_1)_0$ represents the state of the system in the absence of the external flux, that is, when $J_2 = 0$. Equation (8.6) now becomes

$$X_1 J_1 = (X_1 J_1)_0 - X_2 J_2 \tag{8.7}$$

For simplification, assume that the force X_1 is not influenced by change in external flux (i.e., $X_1 = (X_1)_0$). In the case of a fatigue problem, for example, this would imply that the external flux does not affect the stress distribution in the specimen. Therefore, Equation (8.7) yields

$$J_1 = (J_1)_0 - \frac{X_2}{X_1} J_2 \tag{8.8}$$

Equation (8.8) reveals that by increasing the external flux J_2, the flux J_1 decreases which, in turn, results in a decrease in the rate of fatigue damage. Note that the two processes (X_1, J_1) and (X_2, J_2) in Equation (8.8) can be any coupled dissipative processes.

Figure 8.1 shows the variation of J_1 as a function of J_2. This figure shows that, theoretically, by increasing J_2 the fatigue damage can be retarded and even eliminated. While it is practically impossible to eliminate the fatigue damage, this simplified analysis illustrates the possibility of reducing the rate of fatigue damage by the use of an external element such as electric current, magnetic field, or surface cooling.

In Sections 8.2, 8.3, and 8.4, we review some of the pertinent literature on the effect of electric current, magnetic field, and surface cooling on fatigue life, respectively. Concluding remarks on self-organization and complexity are given in Section 8.5.

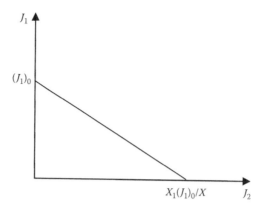

FIGURE 8.1 Effect of external flux on the self-organized system.

8.2 EFFECT OF ELECTRIC CURRENT ON FATIGUE LIFE

Conrad et al. (1991) studied the effect of relatively high-density electric current pulses on the fatigue life of copper. Their experiments involved the rotating-bending fatigue of cold-drawn tough pitch copper rod of 12.5 mm in diameter with a stated purity of 99.9%. The current density was about 1.3×10^4 A/cm^2 for a period of 100 μs with 2 current pulses per second. The frequency of the tests was 50 Hz and the tests were performed at 300 K. They showed that the fatigue can be improved by a factor of 1.3–3 by supplying electric current pulses. Figure 8.2 compares the results of the application of stress-cycle with and without application of electrical pulse for two different copper grain sizes. While the effect of electric current pulses of both grain sizes is noticeable, the procedure is more effective for lower stress amplitudes.

Conrad et al. (1991) postulated that by applying the electric current, the number of cycles required for initiation of the microcracks increases, thus resulting in an increased fatigue life and a decreased tendency for intergranular cracking. Physically, they attributed this effect to the increase in homogenization of slip, that is, the decrease in spacing and width of persistent slip bands (Cao 1989; Conrad et al. 1991). They suggested that homogenization of slip may be caused by interaction between electrons and dislocations. From a thermodynamic viewpoint, the homogenization of slip is analogous to the formation of the dissipative structures as a consequence of self-organization. There are other pertinent works on the effect of electric current of fatigue life of metals; see, for example, Karpenko et al. (1976), Abd El Latif (1979), and Bezborodko (1984).

It is to be mentioned that the work of Abd El Latif (1979) on the effect of pre-application of high-density ac on fatigue life of mild steel specimens showed significant reduction of endurance limit. That is, in his experiments, by applying current prior to the test, the fatigue life of the steel specimen decreased, and that the higher the imposed ac, the lower the fatigue life. However, the notable difference is that Abd El Latif (1979) applied ac and mechanical load separately. Therefore, his testing procedure was completely different from the work of Conrad et al. (1991) where both current and load were applied simultaneously. Abdellatif attributed the unfavorable effect of ac on the fatigue life to the Joule heating and subsequent induced thermal stress, which in turn degrades the material.

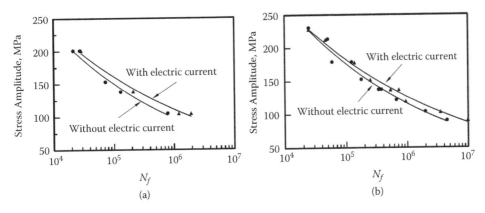

FIGURE 8.2 Effect of electric current pulses on fatigue life in copper with purity of 99.9%. (a) Material with grain size = 45 μm. (b) Material with grain size = 30 μm. (Reproduced from Conrad, H., White, J., Cao, W.D., Lu, X.P., and Sprecher, A.F. et al., *Mater. Sci. Eng. A* 145, 1–12, 1991.)

8.3   EFFECT OF MAGNETIC FIELD ON FATIGUE LIFE

Another technique for enhancement of fatigue life is by the application of a magnetic field on a specimen undergoing fatigue load. Experimental work of Yong, Huacan, and Xiaodong (1993) showed that by applying an alternating magnetic field, the life of a specimen subjected to cyclic fatigue substantially increased. Their experiments were conducted using a uniaxial fatigue testing apparatus with smooth rod specimens made of A3 Steel. The load ratio was $R = 0.01$ and frequency was $f = 25$ Hz. Two series of tests were carried out with and without a magnetic field. Specimens were placed inside a solenoid and a magnetic field was formed by passing electrical current through the solenoid. The experiments were repeated three times at each stress level (see Table 8.1) (Yong et al. 1993).

It was observed that the fatigue life of the specimens was significantly improved under the application of magnetic field. The percentage of increase in fatigue life reported was 269%. While no thermodynamic analysis was presented, Yong, Huacan, and Xiaodong (1993) attributed this effect to the formation of dissipative structures and self-organization during the course of the fatigue. Additional published works on the effect of magnetic field to fatigue life of metals can be found in Dieter (1961); Velez (1997); Bhat, Muju, and Mazumdar (1993); Fahmy et al. (1998); Celik et al. (2005); Aleksandrov and Shakhov (2005); Greger, Kander, and Kocich (2008); and Long et al. (2010).

8.4   EFFECT OF ENVIRONMENT (SURFACE COOLING) ON FATIGUE LIFE

Turning our attention now to the fatigue problem with surface cooling, let us first review some of the pertinent literature that demonstrates environmental influences on life span of materials subjected to fatigue.

The study of environmental effects on fatigue of metals has been of interest for decades since engineering applications sometimes require operation in harsh environments. Effects such as elevated or cryogenic temperatures (Bensely et al. 2007; Franco, Grac, and Silva 2008; Fredj and Sidhom 2006; Mendez 1999; Wiencek, Czarski, and Skowronek 2001),

TABLE 8.1

Effect of the Magnetic Field on the Fatigue Life of A3 Steel

	Without Magnetic Field			With Magnetic Field		
Specimen Number	1	2	3	1	2	3
Cycle to failure, N_f	77880	77840	94050	284610	296980	33950
Average N_f		83256			307380	
Increase in life, %				269%		

corrosive environment (Azevedo and Santos 2003; Bolotin and Shipkov 2001; Creager and Paris 1967; Huang and Shaw 1995; Jaske et al. 1978; Michalska, Sozanska, and Hetmanczyk 2009; Nordmark and Fricke 1978; Radon, Branco, and Culver 1976; Tu and Seth 1978), and vacuum or pressurized environment (Duquette and Gell 1971; Lee, Vasudevan, and Glinka 2009; Mendez and Demulsant 1996; Shen, Podlaseck, and Kramer 1966) can have a pronounced influence on fatigue life of metals.

Environmental conditions can affect the fatigue crack growth rate as it is influenced by surface condition (Parker and Parker 1991; Tian et al. 2006; Tobushi et al. 1997) and, subsequently, can affect the fatigue life. For example, Tobushi et al. studied the influence of air and water atmosphere, temperature, strain amplitude, and rotational speed on the fatigue life of a shape-memory alloy wire subjected to rotating-bending fatigue. They showed that in low-cycle fatigue below 10^4 cycles, the fatigue life in water is longer than that in air. Water was capable of maintaining the wire temperature at a constant level so that the yield stress was constant. But, in the region of high-cycle fatigue above 10^5 cycles, the fatigue life in water was shorter than that in air, mainly due to corrosion effect.

Hirano et al. (2003) investigated the effects of water flow rate on fatigue life of carbon steel in simulated light-water reactor (LWR) environment. They tested Carbon Steel at 289°C for various dissolved oxygen contents (DO) at strain rates of 0.4, 0.01, and 0.001 percent per second (%/s). Their experimental results showed that at the strain rate of 0.01%/s, the fatigue life increased with increasing the flow rate under all DO conditions. Specifically, they reported that the fatigue life at a 7 m/s flow rate was about three times longer than that at a 0.3 m/s flow rate. This increase in fatigue life was attributed to increase in the crack initiation life and small-crack propagation life.

8.5   SELF-ORGANIZATION AND COMPLEXITY

Generally, it is possible to draw an important conclusion that the process of self-organization during fatigue is possible if one or more independent processes, besides plastic deformation itself, are affecting the system. That is, to initiate self-organization and for dissipative structures to form, at least two dissipative processes are required. For example, in the above-mentioned cases, the primary process is plastic deformation while a supplementary process (electric current, magnetic field, or surface cooling) is needed for initiation of the self-organization.

Gershman and Bushe (2006) discuss how by increasing the number of independent processes in a system, the possibility of initiation of self-organization increases. Increasing the number of interacting processes in the system leads to a more complex system.

TABLE 8.2
Effect of External Elements on Fatigue Life

External Element	Life without External Element, N_f	Life with External Element, N_f	Increase in Life %
Conrad et al. (1991) Electrical current Copper (purity 99.9%)	Low-cycle: 5.2×10^4 High-cycle: 7.4×10^5	Low-cycle: 6.9×10^4 High-cycle: 1.5×10^6	Low-cycle: 33 High-cycle: 112
Yong et al. (1993) Magnetic field A3 Steel	Intermediate-cycles: 8.3×10^4	Intermediate-cycles: 3×10^5	Intermediate-cycles: 260

Fox-Rabinovich et al. (2010) theoretically studied the probability of the occurrence of self-organization under complex conditions. They postulate that the probability, P, of losing stability of a complex system with N simultaneous processes can be evaluated from $P = 1-1/(2N)$. Therefore, by increasing the number of processes, N, the system's complexity increases, which in turn leads to a probable initiation of self-organization. Based on this finding, simultaneous application of the external elements mentioned earlier (electric current, magnetic field, and surface cooling) can increase the possibility of formation of dissipative structure during fatigue process and lead to enhanced fatigue life.

REFERENCES

Abd El Latif, A.K. 1979. Fatigue behaviour from application of a.c. *Eng. Fract. Mech.* 12, 449–454.

Aleksandrov, P.A. and Shakhov, M.N. 2005. Effect of magnetic field on low-cycle fatigue. *Doklady Phys.* 50, 63–65.

Amiri, M. and Khonsari, M.M. 2012. Effect of surface cooling on fatigue life improvement. *Int. J. Solids Struct.* (submitted.)

Azevedo, C.R.F. and Dos Santos, A.P. 2003. Environmental effects during fatigue testing: Fractographic observation of commercially pure titanium plate for cranio-facial fixation. *Eng. Failure Anal.* 10, 431–442.

Bensely, A., Senthilkumar, D., Mohan Lal, D., Nagarajan, G., and Rajadurai, A. 2007. Effect of cryogenic treatment on tensile behavior of case carburized steel-815M17. *Mater. Charact.* 58, 485–491.

Bezborodko, L.G. 1984. Effect of alternating current on the strength of steel undergoing cyclic tensile-compressive and torsional loading. *Inst. Mech., Acad. Sci.* (Ukrainian), 95–98.

Bhat, I.K., Muju, M.K., and Mazumdar, P.K. 1993. Possible effects of magnetic fields in fatigue. *Int. J. Fatigue* 15, 193–197.

Bolotin, V.V. and Shipkov, A.A. 2001. Mechanical aspects of corrosion fatigue and stress corrosion cracking. *Int. J. Solids Struct.* 38, 7297–7318.

Cao, W. 1989. Effect of electric current and field on the behavior of metallic materials. Ph.D. thesis, North Carolina State University.

Celik, A., Yetim, A.F., Alsaran, A., and Karakan, M. 2005. Effect of magnetic treatment on fatigue life of AISI 4140 steel. *Mater. Design*, 26, 700–704.

Conrad, H., White, J., Cao, W.D., Lu, X.P., and Sprecher, A.F. 1991. Effect of electric current pulses on fatigue characteristics of polycrystalline copper. *Mater. Sci. Eng. A* 145, 1–12.

Creager, M. and Paris, P.C. 1967. Elastic field equations for blunt cracks with reference to stress corrosion cracking. *Int. J. Fract. Mech.* 3, 247–252.

Dieter, G.E. 1961. *Mechanical Metallurgy.* New York: McGraw-Hill,.

Duquette, D.J. and Gell, M. 1971. The effect of environment on the mechanism of Stage I fatigue fracture. *Metall. Trans.* 2, 1325–1331.

Fahmy, Y., Hare, T., Tooke, R., and Conrad, H. 1998. Effects of a pulsed magnetic treatment on the fatigue of low carbon steel. *Scripta Materialia* 38, 1355–1358.

Fox-Rabinovich, G.S., Gershman, I.S., Yamamoto, K., Biksa, A., Veldhuis, S.C., Beake, B.D., and Kovalev, A.I. 2010. Self-organization during friction in complex surface engineered tribosystems. *Entropy* 12, 275–288.

Franco, L.A.L., Grac, M.L.A., and Silva, F.S. 2008. Fractography analysis and fatigue of thermoplastic composite laminates at different environmental conditions. *Mater. Sci. Eng. A* 488, 505–513.

Fredj, N.B. and Sidhom, H. 2006. Effects of the cryogenic cooling on the fatigue strength of the AISI 304 stainless steel ground components. *Cryogenics* 46, 439–448.

Gershman, I.S. and Bushe, N.A. 2006. Elements of thermodynamics and self-organization during friction. In *Self Organization during Friction: Advanced Surface-Engineered Materials and Systems Design,* G.S. Fox-Rabinovich and G.E. Totten, ed., Boca Raton, FL: CRC Press, Taylor & Francis.

Glansdorff, P. and Prigogine, I. 1971. *Thermodynamic Theory of Structure, Stability and Fluctuations.* New York: John Wiley & Sons, Inc.

Glazov, M., Llantes, L.M., and Laird, C. 1995. Self-organized dislocation structures (SODS) in fatigued metals. *Phys. Stat. Sol. (a)* 149, 297–321.

Greger, M., Kander, L., and Kocich, R. 2008. Structure and low cycle fatigue of steel AISI 316 after ECAP. *Archi. Mater. Sci. Eng.* 31, 41–44.

Henaff, G., Odemer, G., and Tonneau-Morel, A. 2007. Environmentally-assisted fatigue crack growth mechanisms in advanced materials for aerospace applications. *Int. J. Fatigue* 29, 1927–1940.

Hirano, A., Yamamoto, M., Sakaguchi, K., Shoji, T., and Ida, K. 2003. Effects of water flow rate on fatigue life of carbon steel in simulated LWR environment under low strain rate conditions. *J. Pres. Ves. Tech.* 125, 52–58.

Huang, H. and Shaw, W.J.D. 1995. Effect of cold working on the fracture characteristics of mild steel exposed to a sour gas environment. *Mater. Charact.* 34, 43–50.

Jaske, C.E., Broek, D., Slater, J.E., and Anderson, W.E. 1978. Corrosion fatigue of structural steels in seawater and for offshore applications. *Corrosion Fatigue Tech.,* ASTM STP 642, 19–47.

Kabaldin, Y.G. and Murav'yev, S.N. 2007. Information models of self-organization and fatigue damage of metallic materials. *Russian Eng. Res.* 27, 512–518.

Karpenko, G.V., Kuzin, O.A., Tkachev, V.I., and Rudenko, V.P. 1976. The effect of electric current on the low-cycle fatigue of steels. *Dokl. Akad. Nauk SSSR* 227, 85–86.

Kubin, L.P. and Canova, G. 1992. The modeling of dislocation patterns. *Scripta Metall.* 27, 957–962.

Lee, E.U., Vasudevan, A.K., and Glinka, G. 2009. Environmental effects on low cycle fatigue of 2024-T351 and 7075-T651 aluminum alloys. *Int. J. Fatigue* 31, 1938–1942.

Long, L.Z., Yun, H.H., You, F.T., and San, X.X. 2010. Non-equilibrium statistical theory about microscopic fatigue cracks of metal in magnetic field. *Chin. Phys. B* 19, 108103, 1–6.

Mendez, J. 1999. On the effects of temperature and environment on fatigue damage processes in Ti alloys and in stainless steel. *Mater. Sci. Eng. A* 263, 187–192.

Mendez, J. and Demulsant, X. 1996. Influence of environment on low cycle fatigue damage in Ti-6Al-4V and Ti 6246 titanium alloys. *Mater. Sci. Eng. A* 219, 202–211.

Michalska, J., Sozanska, M., and Hetmanczyk, M. 2009. Application of quantitative fractography in the assessment of hydrogen damage of duplex stainless steel. *Mater. Charact.* 60, 1100–1106.

Nicolis, G. and Prigogine, I. 1977. *Self-Organization in Nonequilibrium Systems: From Dissipative Structures to Order Through Fluctuations.* New York: John Wiley & Sons, Inc.

Nordmark, G.E. and Fricke, W.G. 1978. Fatigue crack arrest at low stress intensities in a corrosive environment. *J. Test Eval.* 6, 301–303.

Parker, E.R. and Parker, W.J. 1991. Method for reducing the fatigue crack growth rate of cracks in the aluminum alloy fuselage skin of an aircraft structure. U.S. Patent, Patent number: 5071492. http://www.google.com/patents/US5071492.

Pisarenko, G.G., Voinalovich, A.V., and Mailo. A.N. 2009. Evolution of discrete phenomena of inelasticity in aluminum alloy under cyclic loading. *Strength Mater.* 41, 113–117.

Radon, J.C., Branco, C.M., and Culver, L.E. 1976. Crack blunting and arrest in corrosion fatigue of mild steel. *Int. J. Fract.* 12, 467–469.

Shen, H., Podlaseck, S.E., and Kramer, I.R. 1966. Effect of vacuum on the fatigue life of aluminum. *Acta Metall.* 14, 341–346.

Suresh, S. 1998. *Fatigue of Materials*, 2nd ed. New York: Cambridge University Press.

Taniguchi, T., Kaneko, Y., and Hashimoto, S. 2009. ECCI observations of dislocation structures around fatigue cracks in ferritic stainless steel single crystals. *IOP Conf. Series: Mater. Sci. Eng.* 3, doi:10.1088/1757-899X/3/1/012020.

Tian, H., Liaw, P.K., Fielden, D.E., Jiang, L., Yang, B., Brooks, C.R., Brotherton, M.D., Wang, H., Strizak, J.P, and Mansur, L.K. 2006. Effects of frequency on fatigue behavior of type 316 low-carbon, nitrogen-added stainless steel in air and mercury for the spallation neutron source. *Metall. Mater. Trans. A* 37, 163–173.

Tobushi, H., Hachisuka, T., Yamada, S., and Lin, P.H. 1997. Rotating-bending fatigue of a TiNi shape-memory alloy wire, *Mech. Mater.* 26, 35–42.

Tu, L.K.L. and Seth, B.B. 1978. Threshold corrosion fatigue crack growth in steels. *J. Test Eval.* 6, 66–74.

Velez, S.B. 1997. The effect of residual magnetic field on the fatigue crack propagation of AerMet 100 alloy steel. Master's thesis, University of Puerto Rico.

Wiencek, K., Czarski, A., and Skowronek, T. 2001. Fractal characterization of fractured surfaces of a steel containing dispersed Fe_3C carbide phase. *Mater. Charact.* 46, 235–238.

Yang, Y., Zheng, H.G., Shi, Z.J., and Zhang, Q.M. 2011. Effect of orientation on self-organization of shear bands in 7075 aluminum alloy. *Mater. Sci. Eng. A* 528, 2446–2453.

Yong, Z., Huacan, F., and Xiaodong. F. 1993. Slowing down metal fatigue damage with a magnetic field. *Eng. Fract. Mech.* 46, 347–352.

9 Entropic Fatigue
In Search for Applications

In the preceding chapters, two major concerns in analysis and simulation of a fatigue process, that is, prediction and prevention of fatigue failure, were discussed in detail. We reviewed a variety of fatigue assessment approaches such as traditional stress (strain) methods, energy methods, and also the state-of-the-art thermography technique, which has now facilitated the development of fatigue science to a great extent. We learned that an approach firmly grounded on the principle of irreversible thermodynamics can provide a useful link between entropy and fatigue in a manner that can be used in practical applications. To this end, the predictive capabilities of an entropic approach in fatigue problems with different laboratory specimen sizes, frequency, loading amplitude, and stress states were discussed. Yet a question arises as to how the entropic approach can best be applied at the design stage. In this chapter, we focus our attention on the broader application of an entropic approach to fatigue and implementation of predictive methodologies.

The outline of this chapter is as follows. We begin by demonstrating the application of an entropic-based fatigue approach to variable-loading amplitude. Next, a systematic methodology for accelerated fatigue testing based on the entropic approach is presented followed by a discussion of its practical benefits.

9.1 APPLICATION TO VARIABLE-LOADING AMPLITUDE AND STRUCTURAL HEALTH MONITORING

As discussed in Section 4.1, most practical applications experience a complex spectrum of loading histories with variable amplitudes. There, the linear fatigue damage hypothesis known as Miner's rule was introduced as a powerful and widely used tool for predicting fatigue life under variable-loading amplitude. Let us now examine the application of the entropic approach to this problem.

Referring to Section 6.2, it was shown that subject to constant stress, the cumulative entropy generation increases monotonically to a finite value known as fracture fatigue entropy (FFE). Regarding variable stress amplitude, studies by Naderi and Khonsari (2011) demonstrated that the hypothesis of constant entropy gain at fracture failure still remains valid in low to intermediate-cycle fatigue of metals. They experimentally demonstrated during a variable-amplitude test, that the cumulative entropy generation at each stress level sums up to FFE. That is, if γ_f represents entropy gain at fracture (FFE) in a constant stress amplitude test and γ_1, γ_2, γ_3, ...γ_n denote entropy accumulation during n load sequence in a variable stress amplitude test, then:

$$\gamma_1 + \gamma_2 + \gamma_3 + \cdots + \gamma_n = \gamma_f \tag{9.1}$$

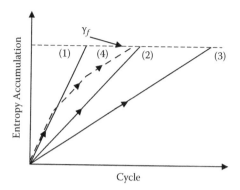

FIGURE 9.1 Entropy accumulations in variable amplitude loading fatigue.

Equation (9.1) can be represented in the form of

$$\sum_{i=1}^{n} \frac{\gamma_i}{\gamma_f} = 1 \tag{9.2}$$

This fact is schematically shown in Figure 9.1. In this figure, paths (1), (2), and (3) correspond to the evolution of cumulative entropy during constant stress amplitudes $\sigma_1 > \sigma_2 > \sigma_3$. See Chapter 6, Figure 6.4 for more detail. Path (4) represents evolution of cumulative entropy during a variable-amplitude test performed sequentially at σ_1, σ_2, and σ_3 for N_1, N_2, and N_3 number of cycles, respectively. This figure shows that regardless of the loading sequence, the accumulation of the entropy generation sums up to γ_f at fracture. In the following example, we show how this fact results in derivation of Miner's rule taking into account cumulative damage (Amiri and Khonsari 2012).

Example 9.1

Consider a component that undergoes a variable-loading amplitude test at stress levels σ_1, ..., σ_n. Let us introduce a damage variable, D, defined as the ratio of the accumulation of entropy generation divided by the fracture fatigue entropy, viz.,

$$D = \frac{\gamma_1 + \cdots + \gamma_n}{\gamma_f} \tag{9.3}$$

where γ_1, ..., γ_n are accumulation of entropy generations at stress levels σ_1, ..., σ_n, respectively. Employ Equation (6.10) to arrive at an equation that represents Miner's cumulative damage law.

Solution

Recalling Equation (6.10), we can write

$$\gamma_f = (\dot{\gamma})_1 (N_f)_1 = \ldots = (\dot{\gamma})_n (N_f)_n \tag{9.4}$$

where $(\dot{\gamma})_1, \ldots, (\dot{\gamma})_n$ denote entropy generations at stress levels σ_1, ..., σ_n, respectively. Note that Equation (9.4) is derived assuming that the stress amplitude is constant. For the case of variable-loading amplitude, the following equations can be deduced from Equation (9.4):

$$\begin{cases} \gamma_1 = (\dot{\gamma})_1 N_1 \\ \quad \vdots \\ \gamma_n = (\dot{\gamma})_n N_n \end{cases} \tag{9.5}$$

where N_1, ..., N_n are the number of cycles amassed at stress levels σ_1, ..., σ_n, respectively. Substituting Equations (9.4) and (9.5) into Equation (9.3), yields

$$D = \frac{(\dot{\gamma})_1 N_1}{(\dot{\gamma})_1 N_{f,1}} + \cdots + \frac{(\dot{\gamma})_n N_n}{(\dot{\gamma})_n N_{f,n}} = \frac{N_1}{N_{f,1}} + \cdots + \frac{N_n}{N_{f,n}} = \sum_{i=1}^{n} \frac{N_i}{N_{f,i}} \tag{9.6}$$

Now, failure occurs when the accumulation of the entropy generation reaches its maximum, that is, γ_f. This condition corresponds to $D = 1$. Therefore, from Equation (9.6) it follows that

$$\sum_{i=1}^{n} \frac{N_i}{N_{f,i}} = 1 \tag{9.7}$$

Equation (9.7) represents the linear fatigue damage hypothesis known as Miner's rule. ▲

9.2 ACCELERATED FATIGUE TESTING

Accelerated testing refers to methodologies that systematically shorten the duration of testing by appropriately expediting the rate of degradation to predict long-term behavior based on short-term laboratory tests.

Full-size fatigue test procedures are expensive, time-consuming, and complex. Even at laboratory scale, fatigue tests often require a long period of testing, especially for high-cycle fatigue. For example, in a typical fatigue test at 10 Hz, 8 weeks would be required to amass 5×10^7 cycles for a single specimen. For a higher number of cycles, experiments may take a long time to complete, and simply increasing the cyclic frequency is not necessarily an appropriate option because it may inversely affect the results. As shown in Figure 9.2, when moving from the field test down to the laboratory environment, test procedure becomes easier, cost of the tests decreases, and extensive permutation of testing becomes more affordable. However, practical significance of the test results necessitates conducting an adequate amount of field testing. Difficulties in addressing failure mechanisms in a fatigue system arise from the inherent complexity of the material behavior in response to the load at multiple length scales (cf. Chapter 4). Larger components and machine parts are observed to have weaker strength against fatigue loading than smaller components and machine parts (Collins 1993). Therefore, fatigue failure is a more serious problem in large size components. Given that initiation and propagation of fatigue crack starts from sites with initial

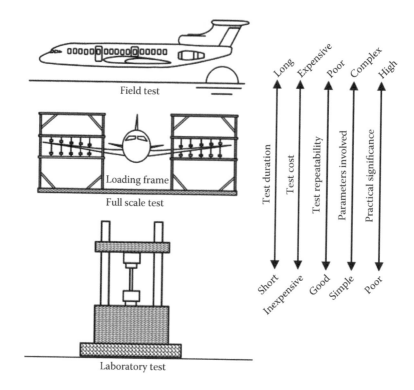

FIGURE 9.2 Advantages and disadvantages of fatigue testing at different scales.

flaws and stress concentration, this is not surprising since there is a greater probability of encountering more critical flaws and stress concentrations in materials with larger volume.

To accelerate a degradation test, researchers recommend increasing stress level, test frequency, or environmental impacts such as temperature and level of corrosivity. (Crowe and Feinberg 2001; Fressinet, Panis, and Bordes 2009; Guerin et al. 2001). As discussed by Crowe and Feinberg, increasing the stress level may cause *nonlinearities,* which, in turn, result in *anomalous* failure. For example, an egg will be hatched under moderate heat, while it will cook if the heat is accelerated (Bryant, Khonsari, and Ling 2008). Thus, increasing the operating factors (stress, frequency, temperature) beyond a certain level may entirely change the failure mode.

Some accelerated fatigue testing models have been proposed by researchers (Allergri and Zhang 2008; Caruso 1996; Caruso and Dasgupta 1998; Farrar et al. 1999). For example, the so-called inverse power law is widely employed in aerospace applications to translate laboratory results to field service life. According to Caruso and Dasgupta, the inverse power law can be described as

$$\frac{N_t}{N_r} = \left(\frac{G_r}{G_t} \right)^{\beta} \tag{9.8}$$

where N_t is the cycles to failure in laboratory accelerated testing, N_r is the fatigue life in real-life application, G represents the G_{rms} value of the load applied to the structure with subscript t and r representing accelerated test and real-life service, respectively (Allergri and Zhang). In their work, Allergri and Zhang enumerate some of the advantages and disadvantages of using the inverse power law (Equation 9.8). One major drawback of using Equation (9.8) to estimate field life from accelerated laboratory data is that it falls short of addressing the effect of variable (or random) stress amplitude, which is commonly observed in practice. We now present a method to implement thermodynamic relations presented in Chapter 3 to develop an accelerated testing scheme based on the concept of thermodynamic forces and flows.

The concept of thermodynamic forces and flows as presented by Equations (3.42) and (3.43) offers a procedure to develop an accelerated fatigue methodology. Let us rewrite these equations here:

$$\frac{d_i s}{dt} = \left(\frac{\partial_i s}{\partial p} \frac{\partial p}{\partial \zeta} \right) \frac{\partial \zeta}{\partial t} = XJ \tag{9.9}$$

$$\frac{dw}{dt} = \left(\frac{\partial w}{\partial p} \frac{\partial p}{\partial \zeta} \right) \frac{\partial \zeta}{\partial t} = YJ \tag{9.10}$$

In these equations, it is assumed that only one dissipative process with a pair of thermodynamic force and flow (X, J) exists. Since the flow rate, $J = d\zeta/dt$, is a common factor in Equation (9.9) and Equation (9.10), one can perform an accelerated fatigue testing by increasing the process rate. Let us define an acceleration factor (AF) as the ratio of the rate of degradation in accelerated testing $(dw/dt)_t$ to the rate of degradation in real-life application $(dw/dt)_r$:

$$AF = \frac{(dw/dt)_t}{(dw/dt)_r} = \frac{(YJ)_t}{(YJ)_r} \tag{9.11}$$

As discussed in Chapter 3, degradation force Y is related to thermodynamic force X via degradation factor B:

$$Y = BX \tag{9.12}$$

Substituting Equation (9.12) into Equation (9.11) yields

$$AF = \frac{B(XJ)_t}{B(XJ)_r} = \frac{(XJ)_t}{(XJ)_r} \tag{9.13}$$

Equation (9.13) suggests that to accelerate a degradation process, one can increase either the thermodynamic force X or the thermodynamic flow J. The following example illustrates the procedure for accelerated testing in fatigue problems.

Example 9.2

Consider a fatigue problem with plastic deformation as a dominant degradation process. Suggest a plan to accelerate testing procedure by a factor of AF.

SOLUTION

Since plastic deformation is the dominant mode of degradation, thermodynamic force and flow can be obtained from (cf. Chapter 3):

$$X = \frac{\sigma}{T}$$

$$J = \dot{\varepsilon}_p \qquad\qquad (9.14)$$

where σ, T, and $\dot{\varepsilon}_p$ are stress amplitude, temperature, and rate of plastic deformation, respectively. Using Equation (9.13) yields:

$$AF = \frac{(XJ)_t}{(XJ)_r} = \frac{(\sigma/T)_t (\dot{\varepsilon}_p)_t}{(\sigma/T)_r (\dot{\varepsilon}_p)_r} \qquad\qquad (9.15)$$

One way to accelerate the fatigue process is to increase the process rate by increasing the rate of plastic strain $\dot{\varepsilon}_p$ as suggested by Equation (9.15). This can be achieved by increasing the test frequency. However, it is to be noted that increasing the frequency (rate of plastic strain) results in an increase in temperature rise. To compensate the unfavorable effect of temperature rise, the thermodynamic force, $X = \sigma/T$, should be adjusted to maintain the same physical failure process. This suggests that one has to choose the stress level, σ, and temperature, T, in such a way so as to maintain an equivalent thermodynamic force. The situation is similar if one decides to perform accelerated testing by increasing the stress level σ. In that case, the plastic strain rate and the temperature should be adjusted to maintain the same physical failure process. ▲

9.3 CONCLUDING REMARKS

Although the notion of entropy and entropy generation for the study of dissipative processes is not new and has a rich history of research in chemistry, biology, material science, and mechanics, its application to practical problems has been realized only in recent years. For example, an interesting study by Silva and Annamalai (2008, 2009) reports to have correlated the biochemical reactions in human metabolism to entropy generation and human lifespan via the first and second laws of thermodynamics. They showed that an individual during his/her life typically accumulates 11,404 kJ/K per kilogram of body mass. This amount predicts 73.78 years of life for an average U.S. male and 81.61 years of life for an average U.S. female. Statistics show an average of 74.63 and 80.36 years of life for males and females, respectively, which are close to the entropic predictions.

The entropic approach to fatigue failure is now gaining momentum in the science community; however, its practical applications have not yet become widespread. We believe that the application of irreversible thermodynamics and specifically entropy generation provides new scientific research opportunities for engineering designs as well as for development of new materials with improved fatigue performance. This concept offers new and exciting research in the field of fatigue fracture for years to come.

REFERENCES

Allegri, G. and Zhang, X. 2008. On the inverse power laws for accelerated random fatigue testing. *Int. J. Fatigue* 30, 967–977.

Amiri, M. and Khonsari, M.M. 2012. On the role of entropy generation in processes involving fatigue. *Entropy* 14, 24–31.

Bryant, M.D., Khonsari, M.M., and Ling, F.F. 2008. On the thermodynamics of degradation. *Proc. R. Soc. A* 464, 2001–2014.

Caruso, H. 1996. An overview of environmental reliability testing, *Proceedings IEEE Annual Reliability and Maintainability Symposium*. ISSN 0149-144X, 102–109.

Caruso, H. and Dasgupta, A. 1998. A fundamental overview of accelerated testing analytic models. *Proceedings Annual Reliability and Maintainability Symposium* 389–393.

Collins, J.A. 1993. *Failure of Materials in Mechanical Design: Analysis, Prediction, Prevention*. New York: John Wiley & Sons, Inc.

Crowe, D. and Feinberg, A. 2001. *Design for Reliability*. Boca Raton, FL: CRC Press.

Farrar, C.R., Duffey, T.A., Cornwell, P.J., and Bement, M.T. 1999. A review of methods for developing accelerated testing criteria. *Proceedings of the 17th International Modal Analysis Conference*, Kissimmee, FL, February 8–11, 1999, 608–614.

Fressinet, M., Panis, M., and Bordes, C. 2009. Accelerated fatigue testing on hydraulic shaker. *25th ICAF Symposium*, Rotterdam, The Netherlands, May 27–29, 2009, 931–938.

Guerin, F., Dumon, B., Hambli, R., and Tebbi, O. 2001. Accelerated testing based on a mechanical-damage model. *Proceedings Annual Reliability and Maintainability Symposium*, IEEE, Philadelphia, PA, January, 22–25, 2001, 372–376.

Naderi, M. and Khonsari, M.M. 2011. Real-time fatigue life monitoring based on thermodynamic entropy. *Structural Health Monitoring* 10, 189–197.

Silva, C.A. and Annamalai, K. 2008. Entropy generation and human aging: Lifespan entropy and effect of physical activity level. *Entropy* 10, 100–123.

Silva, C.A. and Annamalai, K. 2009. Entropy generation and human aging: Lifespan entropy and effect of diet composition and caloric restriction diets. *J. Thermodynamics*, Article ID 186723, 1–10, doi:10.1155/2009/186723.

Index